The cellular events underlying rapid cellular damage are being intensively studied in a number of different organs and all of them are potentially of great importance medically. Examples are the reperfused ischaemic heart, muscular dystrophy, liver damage, malignant hyperthermia and the preservation of kidneys for transplantation. Many hypotheses are extant concerning the biochemical pathways involved in these damage processes and, in particular, the roles of calcium and active oxygen metabolites are of great current interest. There is a considerable literature on both these triggering agents and good evidence for their involvement in the genesis of damage, although little agreement on their precise roles. This volume records the proceedings of a meeting held by the Society for Experimental Biology that attempted to determine whether there are common mechanisms of cellular damage and to explore the ways in which calcium and oxygen radicals may interact to generate the damage. The central problem is: do oxygen radicals modify intracellular calcium levels, with the calcium then triggering the damage, or do changes in calcium fluxes stimulate the generation of oxygen radicals, the radicals then producing the damage?

SOCIETY FOR EXPERIMENTAL BIOLOGY
SEMINAR SERIES : 46

CALCIUM, OXYGEN RADICALS AND CELLULAR DAMAGE

SOCIETY FOR EXPERIMENTAL BIOLOGY SEMINAR SERIES

A series of multi-author volumes developed from seminars held by the Society for Experimental Biology. Each volume serves not only as an introductory review of a specific topic, but also introduces the reader to experimental evidence to support the theories and principles discussed, and points the way to new research.

2. Effects of pollutants on aquatic organisms. *Edited by A.P.M. Lockwood*
6. Neurones without impulses: their significance for vertebrate and invertebrate systems. *Edited by A. Roberts and B.M.H. Bush*
8. Stomatal physiology. *Edited by P.G. Jarvis and T.A. Mansfield*
10. The cell cycle. *Edited by P.C.L. John*
11. Effects of disease on the physiology of the growing plant. *Edited by P.G. Ayres*
12. Biology of the chemotactic response. *Edited by J. M. Lackie and P.C. Williamson*
14. Biological timekeeping. *Edited by J. Brady*
15. The nucleolus. *Edited by E.G. Jordan and C.A. Cullis*
16. Gills. *Edited by D.F. Houlihan, J.C. Rankin and T.J. Shuttleworth*
17. Cellular acclimatisation to environmental change. *Edited by A.R. Cossins and P. Sheterline*
19. Storage carbohydrates in vascular plants. *Edited by D.H. Lewis*
20. The physiology and biochemistry of plant respiration. *Edited by J.M. Palmer*
21. Chloroplast biogenesis. *Edited by R.J. Ellis*
23. The biosynthesis and metabolism of plant hormones. *Edited by A. Crozier and J.R. Hillman*
24. Coordination of motor behaviour. *Edited by B.M.H. Bush and F. Clarac*
25. Cell ageing and cell death. *Edited by I. Davies and D.C. Sigee*
26. The cell division cycle in plants. *Edited by J.A. Bryant and D. Francis*
27. Control of leaf growth. *Edited by N.R. Baker, W.J. Davies and C. Ong*
28. Biochemistry of plant cell walls. *Edited by C.T. Brett and J.R. Hillman*
29. Immunology in plant science. *Edited by T.L. Wang*
30. Root development and function. *Edited by P.J. Gregory, J.V. Lake and D.A. Rose*
31. Plant canopies: their growth, form and function. *Edited by G. Russell, B. Marshall and P.G. Jarvis*
32. Developmental mutants in higher plants. *Edited by H. Thomas and D. Grierson*
33. Neurohormones in invertebrates. *Edited by M. Thorndyke and G. Goldsworthy*
34. Acid toxicity and aquatic animals. *Edited by R. Morris, E.W. Taylor, D.J.A. Brown and J.A. Brown*
35. Division and segregation of organelles. *Edited by S.A. Boffey and D. Lloyd*
36. Biomechanics in evolution. *Edited by J.M.V. Rayner and R. J. Wootton*
37. Techniques in comparative *respiratory physiology*: An experimental approach. *Edited by C.R. Bridges and P.J. Butler*
38. Herbicides and plant metabolism. *Edited by A.D. Dodge*
39. Plants under stress. *Edited by H.G. Jones, T.J. Flowers and M.B. Jones*
40. *In situ* hybridisation: application to developmental biology and medicine. *Edited by N. Harris and D.G. Wilkinson*
41. Physiological strategies for gas exchange and metabolism. *Edited by A.J. Woakes, M.K. Grieshaber and C.R. Bridges*
42. Compartmentation of plant metabolism in non-photosynthetis tissues. *Edited by M.J. Emes*
43. Plant Growth: interactions with nutrition and environment. *Edited by J.R. Porter and D.W. Lawlor*
44. Feeding and the texture of foods. *Edited by J.F.V. Vincent and P.J. Lillford*
45. Endocytosis, excytosis and vesicle traffic in plants. *Edited by G.R. Hawes, J.O.D. Coleman and D.E. Evans*

CALCIUM, OXYGEN RADICALS AND CELLULAR DAMAGE

Edited by

C.J. Duncan

Department of Zoology, University of Liverpool

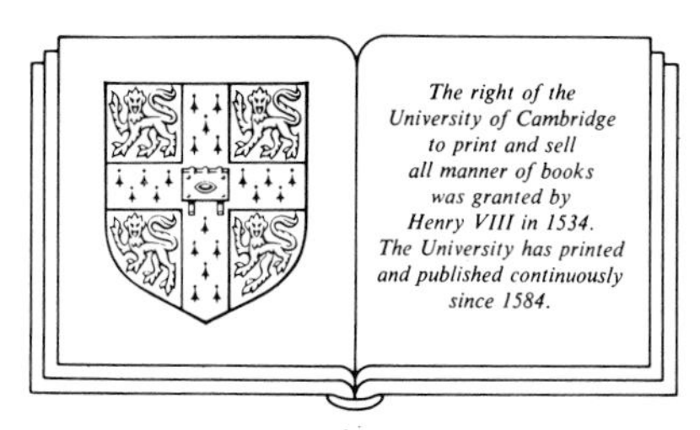

CAMBRIDGE UNIVERSITY PRESS

Cambridge

New York   Port Chester

Melbourne   Sydney

Published by the Press Syndicate of the University of Cambridge
The Pitt Building, Trumpington Street, Cambridge CB2 1RP
40 West 20th Street, New York, NY 10011-4211, USA
10 Stamford Road, Oakleigh, Victoria 3166, Australia

First published 1991

Printed in Great Britain at the University Press, Cambridge

A catalogue record of this book is available from the British Library

Library of Congress cataloguing in publication data

Calcium, oxygen radicals, and cellular damage / edited by C. J. Duncan.
p. cm.—(Society for Experimental Biology seminar series; 46)
'Proceedings of a symposium of the Cell Biology Section of the Society for Experimental Biology, held at Warwick University in March 1990'—Pref.
Includes index.
ISBN 0–521–38068–5
1. Pathology, Molecular–Congresses. 2. Calcium–Toxicology–Congresses. 3. Active oxygen–Toxicology–Congresses.
4. Diseases–Causes and theories of causation–Congresses.
I. Duncan, C. J. (Christopher John) II. Society for Experimental Biology (Great Britain). Cell Biology Section. III. Series: Seminar series (Society for Experimental Biology (Great Britain)); 46.
RB152.C35 1991
616.071–dc20 91–19167 CIP

ISBN 0 521 38068 5 hardback

WV

Contents

List of Contributors ix

Preface xi

Are there common biochemical pathways in cell damage and cell death? 1
C.J. DUNCAN

Free radicals in the pathogenesis of tissue damage 11
M.J. JACKSON

Calcium and signal transduction in oxidative cell damage 17
P. NICOTERA, G.E.N. KASS, S.K. DUDDY AND S. ORRENIUS

Regulation of neutrophil oxidant production 35
S.W. EDWARDS

Reperfusion arrhythmias: role of oxidant stress 77
M.J. SHATTOCK, H. MATSUURA AND D.J. HEARSE

Biochemical pathways that lead to the release of cytosolic proteins in the perfused rat heart 97
C.J. DUNCAN

Malignant hyperthermia: the roles of free radicals and calcium? 115
J.R. ARTHUR AND G.G. DUTHIE

Free radicals, calcium and damage in dystrophic and normal skeletal muscle 139
M.J. JACKSON, A. McARDLE AND R.H.T. EDWARDS

Ultrastructural changes in mitochondria during rapid damage triggered by calcium 149
C.J. DUNCAN AND N. SHAMSADEEN

The importance of oxygen free radicals, iron and calcium in renal ischaemia 165
J.D. GOWER, L.A. COTTERILL AND C.J. GREEN

The Rubicon Hypothesis: a quantal framework for understanding the molecular pathway of cell activation and injury 189
A.K. CAMPBELL

Index 218

Contributors

ARTHUR, J.R.
Division of Biochemical Sciences, Rowett Research Institute, Greenburn Road, Bucksburn, Aberdeen AB2 9SB, UK.
CAMPBELL, A.K.
Department of Medical Biochemistry, University of Wales College of Medicine, Heath Park, Cardiff CF4 4XN, UK.
COTTERILL, L.A.
Section of Surgical Research, Clinical Research Centre, Watford Road, Harrow, Middlesex HA1 3UJ, UK.
DUDDY, S.K.
Department of Toxicology, Karolinska Institutet, Box 60400, S-104 01 Stockholm, Sweden.
DUNCAN, C.J.
Department of Zoology, University of Liverpool, P.O. Box 147, Liverpool L69 3BX, UK.
DUTHIE, G.G.
Division of Biochemical Sciences, Rowett Research Institute, Greenburn Road, Bucksburn, Aberdeen AB2 9SB, UK.
EDWARDS, R.H.T.
Muscle Research Centre, Department of Medicine, University of Liverpool, P.O. Box 147, Liverpool L69 3BX, UK.
EDWARDS, S.W.
Department of Biochemistry, University of Liverpool, P.O. Box 147, Liverpool L69 3BX, UK.
GOWER, J.D.
Section of Surgical Research, Clinical Research Centre, Watford Road, Harrow, Middlesex HA1 3UJ, UK.
GREEN, C.J.
Section of Surgical Research, Clinical Research Centre, Watford Road, Harrow, Middlesex HA1 3UJ, UK.

HEARSE, D.J.
Cardiovascular Research, The Rayne Institute, St. Thomas Hospital, London SE1 7EH, UK.
JACKSON, M.J.
Muscle Research Centre, Department of Medicine, University of Liverpool, P.O. Box 147, Liverpool L69 3BX, UK.
KASS, G.E.N.
Department of Toxicology, Karolinska Institutet, Box 60400, S-104 01 Stockholm, Sweden.
LUNEC, J.
Department of Clinical Chemistry, Wolfson Research Laboratories, Queen Elizabeth Medical Centre, Edgbaston, Birmingham B15 2TH, UK.
McARDLE, A.
Muscle Research Centre, Department of Medicine, University of Liverpool, P.O. Box 147, Liverpool L69 3BX, UK.
MATSUURA, H.
Cardiovascular Research, The Rayne Institute, St. Thomas Hospital, London SE1 7EH, UK.
NICOTERA, P.
Department of Toxicology, Karolinska Institutet, Box 60400, S-104 01 Stockholm, Sweden.
ORRENIUS, S.
Department of Toxicology, Karolinska Institutet, Box 60400, S-104 01 Stockholm, Sweden.
SHATTOCK, M.J.
Cardiovascular Research, The Rayne Institute, St. Thomas Hospital, London SE1 7EH, UK.
SHAMSADEEN, N.S.
Department of Zoology, University of Liverpool, P.O. Box 147, Liverpool L69 3BX, UK.
SLATER, T.F.
Department of Biology and Biochemistry, Brunel University, Uxbridge, Middlesex UB8 3PH, UK.

Preface

This volume represents the proceedings of a Symposium of the Cell Biology section of the Society for Experimental Biology held at Warwick University in March 1990. I am most grateful to Dr Malcolm Jackson for his encouragement and advice during the organisation of a most interesting and successful meeting and to SmithKline Beecham for financial support.

There is currently great interest in the underlying events of rapid cellular damage in a range of different organs, as shown by the chapters in this volume. Calcium ions and oxygen radicals have been implicated in many suggestions concerning the genesis of cellular damage, and the papers in this symposium explored the possibility that these two mechanisms were interrelated. Does intracellular calcium trigger the production of oxygen radicals; or is the main effect of oxygen radicals to damage calcium storage sites and thereby alter calcium homeostasis? Are there common pathways in cellular damage?

C.J. Duncan
Liverpool

C.J. DUNCAN

Are there common biochemical pathways in cell damage and cell death?

The underlying events during rapid cellular damage have been studied in a variety of cells, particularly kidney, hepatocytes and muscle cells, although the initial lesions are very different, being genetic (e.g. Duchenne muscular dystrophy, see Jackson, McArdle & Edwards, this volume; or malignant hyperthermia, see Arthur & Duthie, this volume), or the response to toxic agents, or the result of endocrine dysfunction or a failure in metabolism. It has been suggested that there may be final common pathways, or that there may be a common central trigger. In particular, Ca^{2+} has been proposed as having a major role in initiating and regulating these events in cell death and toxic cell killing (Duncan, 1978; Schanne *et al.*, 1979; Farber, 1981; Trump, Berezesky & Osornio-Vargas, 1981; Nayler, 1983; Orrenius *et al.*, 1989) and consequently these studies have concentrated on the ways in which the initial lesion may result in changes in the intracellular concentration of free Ca^{2+} ($[Ca^{2+}]_i$) and on the pathways that may be activated by a rise in $[Ca^{2+}]_i$. For example, the missing gene product (dystrophin) in Duchenne muscular dystrophy and in *mdx* mice has been identified as a sub-sarcolemma cytoskeletal protein and it remains to be explained how its absence leads to the rise in $[Ca^{2+}]_i$ that has been measured, and to the breakdown of the sarcolemma myofilament apparatus (see Duncan, 1989*a*). There is little doubt that a rise in $[Ca^{2+}]_i$ can precipitate rapid cell damage; it can be produced experimentally by exposure of cells to the divalent cation ionophore A23187 and an elevated $[Ca^{2+}]_i$ has been measured in such conditions as malignant hyperthermia (Lopez *et al.*, 1985).

More recently, interest has focussed on the role of oxygen radicals and active oxygen metabolites in causing cellular damage, and the chemistry and interrelationships of these different species of oxygen radicals are summarised by Jackson (this volume) and in Halliwell & Gutteridge (1984). The neutrophil is a cell that is specialised for the production of superoxides within cytoplasmic vacuoles, thereby effecting the destruction of engulfed bacteria (Edwards, this volume) and illustrating the potency of oxygen radicals and the importance of protective enzymes, such as superoxide dismutase, catalase and glutathione peroxidase. A number of workers have attempted to prevent or ameliorate cell damage by perfusion of hearts or kidneys (Gower, Cotterill & Green, this volume) with these enzymes or with other oxygen radical scavengers; a detailed case has already been made for a major role for oxygen radicals in the damage of the mammalian heart (Hess & Manson, 1984).

This volume aims to examine the roles of Ca^{2+} and oxygen radicals in the genesis of cellular damage in a range of tissues, particularly the ways in which these two mechanisms interact, as in the perfused heart, where they appear to act as simultaneous and interacting triggers (Hearse & Tosaki, 1988). Does Ca^{2+} initiate the production of oxygen radicals or do the latter cause the breakdown in Ca^{2+}-homeostasis and a rise in $[Ca^{2+}]_i$, as has been suggested in hepatocytes (Nicotera *et al.*, this volume)? Alternatively, cellular damage may prove to be the consequence of a web of interacting pathways, as in redox cycling (Kappus, 1986) and in the perfused kidney (Gower *et al.*, this volume). Nevertheless, comparative studies reveal common underlying events; for example, halothane anaesthesia triggers a rise in $[Ca^{2+}]_i$ and severe skeletal muscle damage in the genetically-determined disease malignant hyperthermia (Arthur & Duthie, this volume), but clinical protection is provided by Dantrolene sodium which prevents Ca^{2+} release from the sarcoplasmic reticulum (SR). Halothane or carbon tetrachloride cause hepatic dysfunction via an alteration in Ca^{2+} homeostasis (Zucker, Diamond & Berman, 1982; Farrell *et al.*, 1988) and, interestingly, the hepatotoxicity caused by carbon tetrachloride is also decreased by Dantrolene sodium. Finally, halothane also activates protein kinase C and the generation of superoxides in neutrophils (Tsuchiya *et al.*, 1988).

Ca^{2+} as an intracellular trigger for cellular damage

The systems that control $[Ca^{2+}]_i$ are central to an understanding of the different ways that a rise in $[Ca^{2+}]_i$ may be brought about: (i) release of Ca^{2+} from mitochondria; (ii) release of Ca^{2+} from the endoplasmic reticulum, sarcoplasic reticulum and other intracellular storage and binding sites; (iii) increased Ca^{2+} influx through the plasma membrane; (iv) inhibition of Ca^{2+} efflux from the cell; (v) inhibition of Ca^{2+} uptake at intracellular stores. Some agents may have interacting effects on different sites: A23187 (which causes damage in a variety of cells) promotes both Ca^{2+} entry and Ca^{2+} release from intracellular sites, and metabolic inhibitors or uncouplers may cause both release of mitochondrial Ca^{2+} and the depletion of high energy stores which leads to the failure of active

Ca^{2+} efflux and uptake. An example is the role of intracellular Ca^{2+} in cyanide-induced neurotoxicity in which some protection is provided by pretreatment with the Ca^{2+}-channel blocker diltiazem (Johnson, Meisenheimer & Isom, 1986). Carbon tetrachloride promotes Ca^{2+} release from the endoplasmic reticulum and an enhanced Ca^{2+} influx through the plasma membrane (Orrenius *et al.*, 1989). Hypoxia or anoxia usually operate (particularly in cells that are actively metabolising, as in cardiac muscle) via a failure in oxidative phosphorylation and a consequent fall in ATP supply, thereby raising $[Ca^{2+}]_i$. In this way, hypoxia potentiates carbon tetrachloride hepatoxicity (Shen, Garry & Anders, 1982) and acts synergistically in initiating the breakdown of the sarcolemma in the oxygen paradox of the mammalian heart. The rise in $[Ca^{2+}]_i$ produced by some experimental regimes is achieved only after a series of physiological events; this is exemplified by the Ca^{2+}-paradox of the perfused mammalian heart (described later in this volume by Duncan) and by the myopathy of skeletal muscle produced by the administration of diisopropylfluorophosphate when cholinesterases at the neuromuscular junction are inactivated, so causing a prolongation of the lifetime of the transmitter acetylcholine, a consequent hyperactivation of acetylcholine receptors and an exacerbation of Ca^{2+} entry, with myofilament damage that is confined to the endplate region (Leonard & Salpeter, 1979).

Potential pathways activated by a rise in $[Ca^{2+}]_i$

A rise in $[Ca^{2+}]_i$ can potentially activate a number of biochemical pathways that could culminate in cell damage and cell death and these pathways could interact in a complex web of events (see Gower *et al.*, this volume). In skeletal and cardiac muscle cells damage may result in the destruction of the organisation of the myofilament apparatus and also the breakdown of the sarcolemma, and these are the result of two separate pathways which may be activated independently (Duncan & Jackson, 1987).

Phospholipase A_2 (PLA_2)

Ca^{2+} is required for full activation of PLA_2 and consequently this enzyme has long been considered a candidate for initiating cellular damage mechanisms. It is the regulatory enzyme for the eicosanoid cascade and important products of its activity are lysolecithin and arachidonic acid which acts as the substrate for the cycloxygenase enzymes generating prostaglandins (PG) and lipoxygenase enzymes generating leukotrienes. A role for PGE_2 and $PGE_{2\alpha}$ has been proposed in accelerating muscle protein turnover in fever, trauma and injury (Rodemann & Goldberg,

1982), in Ca^{2+}-stimulated protein degradation in muscle (Rodemann, Waxman & Goldberg, 1982) and in stimulating intralysosomal proteolysis by increasing the production of $PGE_{2\alpha}$ (Baracos *et al.*, 1983). In support of these proposals are the reports of the release of prostaglandins and leukotrienes during Ca^{2+}-stimulated damage in skeletal and cardiac muscle cells (Coker *et al.*, 1981; Karmazyn, 1987).

Active oxygen metabolites can be produced during the eicosanoid cascade and these may be responsible for causing cell damage in muscle cells (see Jackson *et al.*, this volume). However, although PLA_2 and lipoxygenase inhibitors will prevent the release of cytosolic proteins, in skeletal muscle they do not protect against myofilament damage (Duncan & Jackson, 1987; Duncan, 1988*a*).

Alternatively, the arachidonic acid produced may activate a regulatory protein kinase C (Duncan, 1990) and lysolecithin may cause lysosome labilisation (see below).

Ca^{2+}-activated neutral proteases (CANP)

CANPs are thiol proteases that are inhibited by leupeptin and selectively degrade muscle regulatory proteins, cytoskeletal proteins, α-actinin and, ultimately, myofibrils. CANP is localised in muscle cells at the Z-line and it has been suggested that it has a dominant role in the degradation of the myofilament apparatus in Duchenne muscular dystrophy and in Ca^{2+}-stimulated damage in skeletal muscle (Imahori, 1982). Against this view, leupeptin failed to protect against myofilament damage in intact (Duncan, Smith & Greenaway, 1979) and skinned (Duncan, 1989*b*) muscle fibres. However, recent reports show that leupeptin prevents membrane blebbing in hepatocytes, suggesting that alterations in Ca^{2+} homeostasis and a rise in $[Ca^{2+}]_i$ activate CANP which, in turn, degrades the cytoskeleton underlying the plasmalemma which then becomes greatly distorted (Nicotera *et al.*, 1986). CANP is also present in peripheral nerve and brain; it is able to degrade neurofilament proteins and myelin and may be of importance in producing nerve cell damage and demyelination, as in multiple sclerosis (Sato & Miyatake, 1982; Zimmermann & Schlapfer, 1982).

Changes in intracellular pH

A fall in pH_i has been recorded in muscle cells undergoing severe and very rapid damage in response to a rise in $[Ca^{2+}]_i$ (e.g. in malignant hyperthermia) and it is believed to be the result of the hydrolysis of ATP during contraction and of the exchange of H^+ and Ca^{2+} during Ca^{2+} uptake. Such a reduction in pH_i would favour the action of any acid hydrolases released from lysosomes and it has also been suggested that such falls in pH_i could directly cause cellular damage, although there is no convincing evidence for such an hypothesis.

Mitochondrial dysfunction

Mitochondria undoubtedly take up Ca^{2+} when levels of $[Ca^{2+}]_i$ rise abnormally, and swollen mitochondria with marked ultrastructural changes are a common feature of damage, particularly in muscle cells. Wrogemann & Pena (1976) suggested that the aerobic supply of ATP by such mitochondria is markedly reduced and that cellular damage is the consequence of this failure in basic metabolism. It is unlikely that such a sequence of events could explain the remarkable rapidity of some types of damage, such as the release of cytosolic proteins within 90 s in the Ca^{2+} paradox of the heart (Duncan, this volume), but uptake of Ca^{2+} probably activates a mitochondrial PLA_2 whose activity may be responsible for the dramatic ultrastructural changes in the cristal membranes and apparent division of mitochondria that are evident in muscle cells (Duncan, 1988*b*).

Lysosomal enzymes

Both lysosomal and non-lysosomal protein degradation occurs in hepatocytes; the lysosomal pathway is energy-dependent and probably requires Ca^{2+} for the formation or fusion of autophagic vacuoles with the primary lysosomes (Grinde, 1983). The lysosome membrane is notoriously unstable and many agents have been described as having a labilising action, some of which promote rapid cell damage and death. A fall in pH_i described above would increase the activity of acid hydrolases that are released into the cytoplasm. Such observations have led to the suggestion that lysosomal enzymes have a major role in causing rapid cell damage as distinct from their clearly-established functions in protein turnover (Wildenthal & Crie, 1980) or in the final events of cellular degradation. For example, it has been suggested that the prostaglandin E_2 produced during cell damage stimulates intralysosomal proteolysis (Baracos *et al.*, 1983) and activates the lysosomal apparatus (Rodemann *et al.*, 1982); furthermore, intracellular Ca^{2+} stimulates both superoxide generation and the release of lysosomal enzymes in neutrophils (Smolen, Korchak & Weissmann, 1981).

Agents that cause the labilisation of lysosomes include cations, phospholipase, lysolecithin (the product of phospholipase A_2 activity), proteases and low pH_i (Weissmann, 1965), all of which could be included within the suggested pathways advanced above. Of greater interest are the treatments that apparently promote lysosome labilisation in the intact cell but are ineffective on isolated lysosomes; these include dinitrophenol, high oxygen excess (cf. the oxygen paradox of the mammalian heart, see Shattock, Matsuura & Hearse, this volume), ischaemia, anoxia and shock (Weissmann, 1965), all of which are potent in inducing rapid cell damage, some via a rise in $[Ca^{2+}]_i$. Such findings lead to the speculation that Ca^{2+} could cause the breakdown of lysosomes, perhaps via the production of oxygen radicals (Duncan, Greenaway & Smith, 1980). However, there is no direct evidence for the protection against rapid cell damage by a range of protease inhibitors (presented both singly and in a cocktail), and it is concluded that lysosomal enzymes are not involved in the early and severe events of most examples of toxic cell death and that their role is normally confined to the slower and final events of cellular destruction. Lysosomal labilising agents that apparently act only in the intact cell are probably initiating rapid cellular destruction by other, non-lysosomal routes (discussed in the following chapters) which may be mistaken for lysosomal labilisation or which ultimately cause the release of acid hydrolases during the final stages of degradation.

Conclusions

There is good evidence that Ca^{2+} has a central role in triggering rapid cell damage in a variety of cells and it can potentially activate a number of separate pathways, as described in the following chapters. Oxygen radicals are also implicated in some cases of damage and together these pathways may interact in a complex web of events.

References

Baracos, V., Rodemann, P., Dinarello, C.A. & Goldberg, A.L. (1983). Stimulation of muscle protein degradation and prostaglandin E_2 release by leukocytic pyrogen (interleukin-1). *New England Journal of Medicine* **308**, 553–8.

Coker, S.J., Marshall, R.J., Parratt, J.R. & Zeitlin, I.J. (1981). Does the local myocardial release of prostaglandin E_2 or $F_{2\alpha}$ contribute to the early consequences of acute myocardial ischaemia? *Journal of Molecular and Cellular Cardiology* **13**, 425–34.

Duncan, C.J. (1978). Role of intracellular calcium in promoting muscle damage: a strategy for controlling the dystrophic condition. *Experientia* **34**, 1531–5.

Duncan, C.J. (1988*a*). The role of phospholipase A_2 in calcium-induced damage in cardiac and skeletal muscle. *Cell and Tissue Research* **253**, 457–62.

Duncan, C.J. (1988*b*). Mitochondrial division in animal cells. In *The*

Division and Segregation of Organelles, ed. S.A. Boffey & D. Lloyd, pp. 95–113. Cambridge University Press.

Duncan, C.J. (1989*a*). Dystrophin and the integrity of the sarcolemma in Duchenne muscular dystrophy. *Experientia* **45**, 175–7.

Duncan, C.J. (1989*b*). The mechanisms that produce rapid and specific damage to the myofilaments of amphibian skeletal muscle. *Muscle and Nerve* **12**, 210–18.

Duncan, C. J. (1990). Biochemical events associated with rapid cellular damage during the oxygen- and calcium-paradoxes of the mammalian heart. *Experientia* **46**, 41–8.

Duncan, C.J., Greenaway, H.C. & Smith, J.L. (1980). 2,4-dinitrophenol, lysosomal breakdown and rapid myofilament degradation in vertebrate skeletal muscle. *Naunyn-Schmiedeberg's Arch. Pharmacol.* **315**, 77–82.

Duncan, C.J. & Jackson, M.J. (1987). Different mechanisms mediate structural changes and intracellular enzyme efflux following damage to skeletal muscle. *Journal of Cell Science* **87**, 183–8.

Duncan, C.J., Smith, J.L. & Greenaway, H.C. (1979). Failure to protect frog skeletal muscle from ionophore-induced damage by the use of the protease inhibitor leupeptin. *Comparative Biochemistry and Physiology* **63**C, 205–7.

Farber, J.L. (1981). The role of calcium in cell death. *Life Science* **29**, 1289–95.

Farrell, G.C., Mahoney, J., Bilous, M. & Frost, L. (1988). Altered hepatic calcium homeostasis in guinea pigs with halothane-induced hepatotoxicity. *Journal of Pharmacology and Experimental Therapeutics* **247**, 751–6.

Grinde, B. (1983). Role of Ca^{2+} for protein turnover in isolated rat hepatocytes. *Biochemical Journal* **216**, 529–36.

Halliwell, B. & Gutteridge, J.M.C. (1984). Oxygen toxicity, oxygen radicals, transition metals and disease. *Biochemical Journal* **219**, 1–14.

Hearse, D.J. & Tosaki, A. (1988). Free radicals and calcium: simultaneous interacting triggers as determinants of vulnerability to reperfusion-induced arrhythmias in the rat heart. *Journal of Molecular and Cellular Cardiology* **20**, 213–23.

Hess, M.L. & Manson, N.H. (1984). Molecular oxygen: friend and foe. *Journal of Molecular and Cellular Cardiology* **16**, 969–85.

Imahori, K. (1982). Calcium-dependent neutral protease: its characterization and regulation. In *Calcium and Cell Function*, ed. W.Y. Cheung, pp. 473–85. New York: Academic Press.

Johnson, J.D., Meisenheimer, T.L. & Isom, G.E. (1986). Cyanide-induced neurotoxicity: role of neuronal calcium. *Toxicology and Applied Pharmacology* **84**, 464–9.

Kappus, H. (1986). Overview of enzyme systems involved in bioreduction of drugs and in redox cycling. *Biochemical Pharmacology* **35**, 1–6.

Karmazyn, M. (1987). Calcium paradox-evoked release of prostacyclin and immunoreactive leukotriene C4 from rat and guinea-pig hearts. Evidence that endogenous prostaglandins inhibit leukotriene biosynthesis. *Journal of Molecular and Cellular Cardiology* **19**, 221–30.

Leonard, J.P. & Salpeter, M.M. (1979). Agonist-induced myopathy at the neuromuscular junction is mediated by calcium. *Journal of Cell Biology* **82**, 811–19.

Lopez, J.R., Alamo, L., Caputo, C., Wikinski, J. & Ledezma, D. (1985). Intracellular ionized calcium concentration in muscles from humans with malignant hyperthermia. *Muscle and Nerve* **8**, 355–8.

Nayler, W.G. (1983). Calcium and cell death. *European Heart Journal* **4**, 33–41.

Nicotera, P., Hartzell, P., Davis, G. & Orrenius, S. (1986). The formation of plasma membrane blebs in hepatocytes exposed to agents that increase cytosolic Ca^{2+} is mediated by the activation of a non-lysosomal proteolytic system. *FEBS Letters* **209**, 139–44.

Orrenius, S., McConkey, D.J., Bellomo, G. & Nicotera, P. (1989). Role of Ca^{2+} in toxic cell killing. *Trends in Pharmacological Science* **10**, 281–5.

Rodemann, H.P. & Goldberg, A.L. (1982). Arachidonic acid, prostaglandin E_2 and $F_{2\alpha}$ influence rates of protein turnover in skeletal and cardiac muscle. *Journal of Biological Chemistry* **257**, 1632–8.

Rodemann, H.P., Waxman, L. & Goldberg, A.L. (1982). The stimulation of protein degradation in muscle by Ca^{2+} is mediated by prostaglandin E_2 and does not require the calcium-activated protease. *Journal of Biological Chemistry* **257**, 8716–23.

Sato, S. & Miyatake, T. (1982). Degradation of myelin basic protein by calcium-activated neutral protease (CANP)-like enzyme in myelin and inhibition by E-64 analogue. *Biomedical Research* **3**, 461–4.

Schanne, F.A.X., Kane, A.B., Young, E.E. & Farber, J.L. (1979). Calcium dependence of toxic cell death: a final common pathway. *Science* **206**, 700–2.

Shen, E.S., Garry, V.F. & Anders, M.W. (1982). Effect of hypoxia on carbon tetrachloride hepatotoxicity. *Biochemistry and Pharmacology* **31**, 3787.

Smolen, J.E., Korchak, H.M. & Weissmann, G. (1981). The roles of extracellular and intracellular calcium in lysosomal enzyme release and superoxide anion generation by human neutrophils. *Biochimica et Biophysica Acta* **677**, 512–20.

Trump, B.F., Berezesky, I.K. & Osornio-Vargas, A.R. (1981). Cell death and the disease process. The role of calcium. In *Cell Death in Biology and Pathology*, ed. I.D. Bowen & R.A. Locksin. London: Chapman and Hall.

Tsuchiya, M., Okimasu, E., Ueda, W., Hirakawa, M. & Utsumi, K. (1988). Halothane, an inhalation anesthetic, activates protein kinase C and superoxide generation by neutrophils. *FEBS Letters* **242**, 101–5.

Weissmann, G. (1965). Lysosomes (concluded). *New England Journal of Medicine* **273**, 1143–9.

Wildenthal, K. & Crie, J.S. (1980). The role of lysosomes and lysosomal enzymes in cardiac protein turnover. *Federal Proceedings* **39**, 37–41.

Wrogemann, K. & Pena, S.D.J. (1976). Mitochondrial calcium overload: a general mechanism for cell-necrosis in muscle diseases. *Lancet* March 27, pp. 672–4.

Zimmermann, U.P. & Schlapfer, W.W. (1982). Characterization of a brain calcium-activated protease that degrades neurofilament proteins. *Biochemistry* **21**, 3977–83.

Zucker, J.R., Diamond, E.M. & Berman, M.C. (1982). Effect of halothane on calcium transport in isolated hepatic endoplasmic reticulum. *British Journal of Anaesthesia* **54**, 981–5.

M.J. JACKSON

Free radicals in the pathogenesis of tissue damage

Free radicals are by definition atoms or groups of atoms containing an unpaired electron. For many years it was felt that such reactive molecules were in the realm of inorganic chemists and had no real relevance to biology and medicine, but in recent years substantial numbers of papers have been published providing evidence of the involvement of free-radical-mediated reactions in various pathological (and some physiological) situations. Perhaps the most completely studied of these processes in the hepatotoxicity which results from carbon tetrachloride exposure (Slater, Cheeseman & Ingold, 1985), while the area which has attracted most attention is the possibility that free radical species mediate the damage to tissues which occurs during ischaemia and reperfusion of tissues, such as the heart following myocardial infarction (McCord, 1985; Ytrehus *et al.*, 1987; Badylak *et al.*, 1987). This chapter will briefly review potential sources of free radicals *in vivo*, the body's protective mechanisms against radicals and the inherent problems in studying these processes during tissue damage.

Cellular sources of free radicals

The presence of an unpaired electron can convey considerable (potentially uncontrolled) reactivity to the free radical and the body therefore generally relies on enzymatic catalysis of reactions which do not involve free radicals. There are however a number of exceptions to this rule in which endogenous production of free radicals occurs during aspects of normal aerobic metabolism. The superoxide anion [$O_2^{\dot{-}}$] is produced by several cell redox systems including xanthine oxidase and membrane-associated NADPH oxidase.

Xanthine oxidase

$$\text{Xanthine} + 2O_2 + H_2O \rightarrow \text{Urate} + 2O_2^{\dot{-}} + 2H^+$$

NADPH oxidase

$$\text{NADPH} + 2O_2 \rightarrow \text{NADP}^+ \; 2O_2^{\dot{-}} + H^+$$

In particular, phagocytic cells have greatly increased oxygen uptake on stimulation and utilise the NADPH oxidase to release large amounts of the superoxide anion into extracellular fluid (Kleblanoff, 1982). In addition Forman & Boveris (1982) have suggested that about 1–4% of the total oxygen taken up by mitochondria may be converted to superoxide and potentially released from the mitochondria. Thus tissues such as muscle which increase their oxygen uptake during exercise might also generate greatly increased amounts of superoxide (Chance, Sies & Boveris, 1979). Superoxide also appears to be produced in xanthine oxidase containing tissues during ischaemia and reperfusion (McCord, 1985).

The superoxide anion is not particularly reactive, but is capable of diffusing throughout relatively large distances in the cell where in the presence of iron or copper a metal-catalysed Haber–Weiss reaction can occur with the formation of the highly reactive hydroxyl radical ($^{\cdot}OH$) (Haber & Weiss, 1934; Halliwell & Gutteridge, 1985).

Iron-catalysed Haber–Weiss reaction

$$2O_2^{\cdot -} + 2H^+ \rightarrow H_2O_2 + O_2$$

$$O_2^{\cdot -} + Fe^{3+} \rightarrow Fe^{2+} + O_2$$

$$Fe^{2+} + H_2O_2 \rightarrow Fe^{3+} + OH^- + {}^{\cdot}OH$$

This hydroxyl free radical rapidly reacts with biological molecules in its immediate vicinity at an essentially diffusion controlled rate (Willson, 1978) and is thought by many workers to be the key radical species in causing tissue damage in a number of diverse pathological conditions (Halliwell & Gutteridge, 1985). It can be seen that in order for this suggestion to be true it is essential that iron or copper are available in tissues or extracellular fluid in a form capable of catalysing the above reaction. Both iron and copper are generally present in a protein-bound form within the body, but the possibility that certain iron complexes are able to participate in hydroxyl radical generating reactions has received considerable attention (Gutteridge, 1987; Aruoma & Halliwell, 1987) although the demonstration that such reactions can occur *in vivo* is problematic and controversial.

The metabolism of arachidonic acid and other polyunsaturated fatty acids by both cyclo-oxygenase and lipoxygenase enzymes also provides another potential source of free radicals. As previously mentioned, radical intermediates occur in both the cyclo-oxygenase and lipoxygenase pathways of metabolism and in addition the peroxides produced may be converted to either peroxy($ROO^{\cdot}$) or alkoxy ($RO^{\cdot}$) radicals on reaction with 'catalytic' iron or copper complexes (Jackson, Jones & Edwards, 1983).

$$Fe^{2+} + ROOH \rightarrow Fe^{3+} + RO^{\cdot} + OH^{-}$$
$$Fe^{3+} + ROOH \rightarrow Fe^{2+} + ROO^{\cdot} + H^{+}$$

These radicals have high reactivity with other polyunsaturated fatty acids and can initiate the process of non-enzymatic lipid peroxidation by which membrane fatty acids can be degraded.

Defence mechanisms against free radicals

The accumulation of the superoxide anion is prevented by enzymes called superoxide dismutases which contain manganese or copper and zinc at their active site. In mammalian cells the manganese-containing enzyme is located in the mitochondria and the copper/zinc-containing enzyme is located in the cytoplasm so that both mitochondrial and cytoplasmic enzymes are protected. The hydrogen peroxide formed from the action of superoxide dismutase is prevented from forming the hydroxyl radical by glutathione peroxidase and catalase and the former enzyme also acts to prevent the accumulation of lipid peroxides in the cytosol (Halliwell & Gutteridge, 1985). In addition vitamin E and β-carotene are lipid-soluble antioxidants preventing radical-mediated breakdown of polyunsaturated fatty acids and vitamin C acts as a reductant in the cell, both regenerating oxidised vitamin E and preventing oxidation of cytosolic thiols.

Together these and other less specific systems combine to provide the cell with substantial protection against the possible deleterious effects of free radical species. However, as has been previously suggested, under certain circumstances substantial over-production of radical species may occur which may overcome those defensive systems and allow reaction of free radicals with cellular components sufficient to result in damage to the cell.

Free radical damage to the cell

Free radical species, particularly the hydroxyl radical, are highly reactive and can react with almost all constituents of the cell. Of particular importance appear to be the reaction of free radicals with membrane lipids (to cause non-enzymatic lipid peroxidation) and with DNA (potentially to cause strand breakage). Reaction of free radicals with proteins, particularly thiol groups of proteins, has received much less attention although this may eventually prove to be an important aspect of radical damage to cells (Halliwell & Gutteridge, 1985). In the light of the preceding discussion it is unsurprising that in experimental situations treatment of tissues with radical-generating agents can cause morphological and functional damage to tissue. For instance, although large intravenous injections of linoleic acid hydoperoxide can damage the aortic intima of rabbits (Yagi *et al.*, 1981) and both hydrogen peroxide and organic peroxides can cause functional changes in erythrocyte membranes (Van der Zee, Dubbelman & van Steverinck, 1985), it does not necessarily follow that these processes are related to any aspects of human pathology. It is therefore important in studies attempting to implicate free radicals in pathological processes to demonstrate increased indices of free radical activity in the situation (disorder or model system) under study *and* whether these and the ensuing damage can be prevented by inhibitors/scavengers of radical reactions. These two key observations have not been reported in many of the pathological situations in which free radicals have been suggested to be implicated. In particular elevated levels of indirect indicators of free radical activity have been described in a large number of human pathological conditions with no evidence that the disease processes are modified by antioxidants, and protective effects of non-specific radical scavengers have been described in many experimental situations where no evidence of elevated free radical activity has been provided. The combination of these two findings is important because either one alone is subject to possible misinterpretation, thus indirect indicators of radical activity are notoriously non-specific and elevated secondary to tissue damage (Halliwell & Gutteridge, 1984) and inhibitors of radical reactions such as vitamin E may have substantial protective effects on cells unrelated to their antioxidant role (Phoenix, Edwards & Jackson, 1989; 1990).

Relationship of radical damage to other processes

Much of the initial work on the possible involvement of radical species in cell damage has concentrated on potential direct major deleterious effects of the radicals, such as the breakdown of membranes via non-enzymatic lipid peroxidation or chain breakage in DNA. It is now becoming recognised that where radicals are playing a part in cell damage it may well be as a component of a more complex damaging process. Thus the induction of hepatotoxicity in carbon tetrachloride poisoning is now recognised to involve the generation of radial species with consequent lipid peroxidation and a failure of calcium homeostasis, both contributing to cell death (Recknagel & Glende, 1989). Conversely the condition of malignant hyperthermia, originally thought to be entirely due to a failure of calcium homeostasis (O'Brien *et al.*, 1990) appears to show evidence of increased free radical activity and partially responds to vitamin E supplementation (Duthie *et al.*, 1990; Duthie & Arthur, 1989). Elucidation of the relative importance and roles of such processes will be an important step in our understanding of the mechanisms underlying cellular damage and in determining optimal therapeutic regimens to protect the tissues.

References

Aruoma, O.I. & Halliwell, B. (1987). Superoxide-dependent and ascorbate-dependent formation of hydroxyl radicals from hydrogen peroxide in the presence of iron. Are lectoferrin and transferrin promoters of hydroxyl-radical generation? *Biochemical Journal* **241**, 273–8.

Badylak, S.F., Simmons, A., Turek, J. & Babbs, C.F. (1987). Protection from reperfusion injury in the isolated rat heart by post-ischaemic desferrixamine and oxypurinol administration. *Cardiovascular Research* **21**, 500–6.

Chance, B., Sies, H. & Boveris, A. (1979). Hydroperoxide metabolism in mammalian organs. *Physiological Reviews* **59**, 527–605.

Duthie, G.G. & Arthur, J.R. (1989). The antioxidant abnormality in the stress susceptible pig: Effect of vitamin E supplementation. *Annals of the New York Academy of Science* **570**, 322–34.

Duthie, G.G., McPhail, D.B., Arthur, J.R., Goodman, B.A. & Morrice, P.C. (1990). Spin trapping of free radicals and lipid peroxidation in microsomal preparations from malignant hyperthermia susceptible pigs. *Free Radical Research Communications* **8**, 93–9.

Forman, H.J. & Boveris, A. (1982). Superoxide radical and hydrogen peroxide in mitochondria. In *Free Radicals in Biology*, ed. W.A. Pryor, pp. 65–90. New York: Academic Press.

Gutteridge, J.M.C. (1987). The role of oxygen radicals in tissue damage and ageing. *Pharmacology Journal* **239**, 401–6.

Haber, F. & Weiss, J. (1934). The catalytic decomposition of hydrogen peroxide by iron salts. *Proceedings of the Royal Society of London* A **747**, 332–51.

Halliwell, B. & Gutteridge, J.M.C. (1984). Lipid peroxidation, oxygen radicals, cell damage and antioxidant therapy. *Lancet* **ii**, 1396–7.

Halliwell, B. & Gutteridge, J.M.C. (1985). In *Free Radicals in Biology and Medicine*. Oxford: Clarendon Press.

Jackson, M.J., Jones, D.A. & Edwards, R.H.T. (1983). Lipid peroxidation of skeletal muscle – an 'in vitro' study. *Bioscience Reports* **3**, 609–19.

Klebanoff, S.J. (1982). Oxygen-dependent cytotoxic mechanisms of phagocytes. In *Advances in Host Defence Mechanisms*, vol. 1, ed. J.I. Gallin & A.S. Fauci, pp. 111–62. New York: Raven Press.

McCord, J.M. (1985). Oxygen derived free radicals in post-ischaemic tissue injury. *New England Journal of Medicine* **312**, 159–63.

O'Brien, P.J., Klip, A., Britt, B.A. & Kalow, B.I. (1990). Malignant hyperthermia susceptibility: biochemical basis for pathogenesis and diagnosis. *Canadian Journal of Veterinary Research* **54**, 83–92.

Phoenix, J., Edwards, R.H.T. & Jackson, M.J. (1989). Inhibition of calcium-induced cytosolic enzyme efflux from skeletal muscle by vitamin E and related compounds. *Biochemical Journal* **287**, 207–13.

Phoenix, J., Edwards, R.H.T. & Jackson, M.J. (1990). Effects of calcium ionophore on vitamin E deficient muscle. *British Journal of Nutrition* **64**, 245–56.

Recknagel, R.O. & Glende, E.A. Jr. (1989). The carbon tetrachloride hepatotoxicity model: Free radicals and calcium homeostasis. In *CRC Handbook of Free Radicals and Antioxidants in Biomedicine*, Vol III, ed. J. Miquel, A.T. Quintanilla & H. Weber. Boca Raton: CRC Press.

Slater, T.F., Cheeseman, K.H. & Ingold, K.U. (1985). Carbon tetrachloride toxicity as a model for studying free-radical mediated liver injury. *Philosophical Transactions of the Royal Society of London* B**311**, 633–6.

Van der Zee, J., Dubbelman, T.M.A.R. & van Steverinck, J. (1985). Peroxide-induced membrane damage in human erythrocytes. *Biochimica et Biophysica Acta* **818**, 38–43.

Willson, R.L. (1978). Free radicals and tissue damage: Mechanistic evidence from radiation studies. In *Biochemical Mechanisms of Liver Injury*, ed. T.F. Slater, pp. 123–224. London: Academic Press.

Yagi, K., Ohkawa, H., Ohistii, N., Yamashita, M. & Nakashima, T. (1981). Lesion of aortic intima caused by intravenous administration of linoleic hydroperoxide. *Journal of Applied Biochemistry* **3**, 58–61.

Ytrehus, K., Gunner, S., Myklebust, R. & Mjos, O.D. (1987). Protection by superoxide dismutase and catalase in the isolated rat heart reperfused after prolonged cardioplegia: a combined study of metabolic, functional and morphometric ultrastructural variables. *Cardiovascular Research* **21**, 492–9.

PIERLUIGI NICOTERA, GEORGE E.N. KASS, STEVEN K. DUDDY
and STEN ORRENIUS

Calcium and signal transduction in oxidative cell damage

Introduction

In aerobically growing eukaryotic cells, ATP synthesis is coupled to the enzymatic reduction of molecular oxygen (O_2) to H_2O via a sequential, four-electron transfer reaction. However, one- and two-electron reduction of O_2 can also occur under physiological conditions in mitochondria and other cellular compartments, generating superoxide anion free radicals ($O_2^{\cdot-}$) and other active oxygen species, including hydrogen peroxide (H_2O_2) and the hydroxyl radical ($^{\cdot}OH$) (Fridovich, 1983). Although several intracellular mechanisms exist that normally prevent the accumulation of active oxygen species during cell metabolism, these protective systems can become overwhelmed and toxic damage to cells may follow. Thus, accumulating evidence implicates oxygen radicals and other oxygen-derived species as causative agents in ageing and a variety of human diseases, including cancer (Cerutti, 1985).

The term 'oxidative stress' is generally applied to conditions in which the intracellular prooxidant–antioxidant ratio favours the former (Sies, 1985). Recent studies by our laboratory and others have examined the effects of oxidative stress on various cellular functions. Experimental systems commonly used to expose various types of cells and tissues to oxidative stress include enzymatic generation of active oxygen species (e.g. using xanthine/xanthine oxidase) (Muehlematter, Larsson & Cerutti, 1988), direct exposure to peroxides or prooxidants (Sies, Brigelius & Graf, 1987) and intracellular generation of active oxygen species by treatment with redox-cycling chemicals such as quinones or bipyridilium compounds (Smith *et al.*, 1985; Orrenius *et al.*, 1986). Notable biochemical consequences of cellular oxidative stress include marked alterations in the regulation of intermediary metabolism (Sies *et al.*, 1987; Sies, 1985), disruption of intracellular thiol and Ca^{2+} homeostasis (Di Monte *et al.*, 1984*a*,*b*; Nicotera *et al.*, 1988), modification of signal transduction pathways (Bellomo, Thor & Orrenius, 1987;

Gopalakrishna & Anderson 1989; Kass, Duddy & Orrenius, 1989*b*) and tumor promotion (Cerutti, 1985).

Role of glutathione and protein thiol modification during oxidative stress

Glutathione (GSH) plays a unique role in the cellular defence against active oxygen species and reactive intermediates. GSH functions both as a reductant in the metabolism of hydrogen peroxide and organic hydroperoxides and as a nucleophile which can conjugate electrophilic molecules (Orrenius & Moldéus, 1984). During glutathione peroxidase-catalysed metabolism of hydroperoxides, GSH serves as an electron donor, and the glutathione disulphide (GSSG) formed in the reaction is subsequently reduced back to GSH by glutathione reductase, at the expense of NADPH. Under conditions of oxidative stress, when the cell must cope with large amounts of H_2O_2 or organic hydroperoxides, the rate of glutathione oxidation exceeds the slower rate of GSSG reduction by glutathione reductase, and GSSG accumulates. To avoid the detrimental effects of increased intracellular levels of GSSG (e.g. formation of mixed disulphides with protein thiols), the cell actively excretes GSSG, which can lead to depletion of the intracellular glutathione pool (Fig. 1).

Although it is now well established that cell killing caused by conditions of oxidative stress is preceded by depletion of intracellular GSH (Nicotera & Orrenius, 1986), the exact relationship between GSH depletion and cell death has not yet been clarified. Sulphydryl groups are, in general, highly reactive; thus, the sulphydryl groups contained in molecules other than GSH, for example cellular protein thiols, may represent critical targets for oxidative-stress-mediated cell killing. Indeed, the generation of conditions of oxidative stress during the metabolism of menadione (2-methyl-1,4-naphthoquinone) in isolated rat hepatocytes results in the loss of protein thiols. This loss of protein thiols follows glutathione depletion and precedes the onset of cell death. Moreover, pretreatments which deplete intracellular GSH (e.g. diethylmaleate), potentiate menadione-induced protein thiol depletion and cytotoxicity (Di Monte *et al.*, 1984*a*).

Oxidative stress disrupts the cytoskeleton

Exposing hepatocytes to menadione to generate conditions of oxidative stress results in an alteration of surface morphology characterised by loss of microvilli and the appearance of multiple blebs on the surface of the hepatocytes (Thor *et al.*, 1982). Many other toxic agents, including

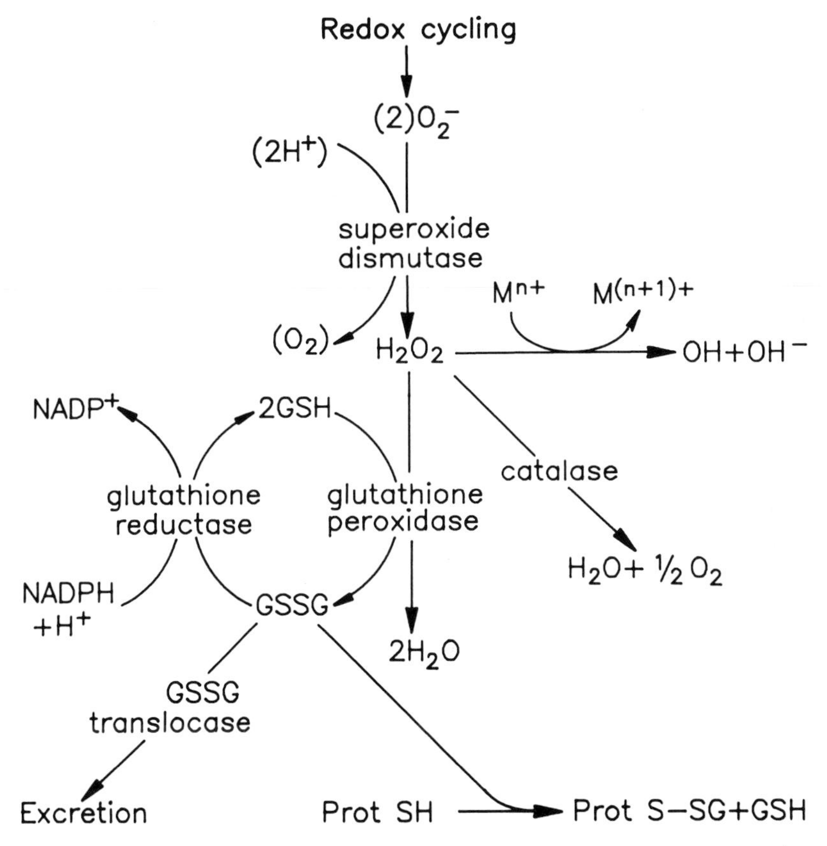

Fig. 1. Formation and metabolism of active oxygen species.

1-methyl-4-phenyl-1,2,3,6-tetrahydropyridine (MPTP) (Kass *et al.*, 1988) and the Ca^{2+} ionophore A23187 (Jewell *et al.*, 1982), cause similar alterations in surface structure, indicating that plasma membrane blebbing is a common event in the progression of toxic injury. Such blebs appear before alterations of the plasma membrane permeability characteristics and seem to be initially reversible.

From the findings that hepatocyte blebbing can also be induced by the microfilament modifying drugs cytochalasin D, phalloidin, and microcystin-LR (Prentki *et al.*, 1979; Eriksson *et al.*, 1989), it appears that the blebbing caused by menadione and other cytotoxic agents may be due to microfilament disruption. The direct relationship between perturbation of the cytoskeleton during oxidative stress and the resulting cell killing is at present unclear. Many cellular processes such as cell division, intracellular transport, cell to cell contact, and the control of cell motility and shape are critically dependent on an intact cytoskeleton; disruption of such processes by oxidative stress would clearly be detrimental to cell survival. Furthermore, blebs have been observed to rupture, compromising the integrity of the plasma membrane.

Mechanistically, perturbation of normal cytoskeletal organization by oxidative stress may result from the disruption of intracellular thiol and Ca^{2+} homeostasis. Thiol oxidation perturbs microfilament organisation through formation of disulphide cross-links between actin molecules (Mirabelli *et al.*, 1988). A change in intracellular Ca^{2+} distribution can affect cytoskeletal structure because calcium ions and cytoskeleton-associated Ca^{2+}-binding proteins play a pivotal role in regulation of the microfilament network (Weeds, 1982). Alternatively, Ca^{2+}-activated catabolic enzymes, such as the calpain family of cytosolic proteases, can cleave cytoskeleton-associated proteins, and thereby cause alterations of the microfilament network and blebbing (Nicotera *et al.*, 1986; Mirabelli *et al.*, 1989). It is likely that both thiol modification and Ca^{2+}-dependent processes (see below) may combine to produce cytoskeletal alterations during oxidative stress, and that this may play a determinant role in the onset of cell killing.

Oxidative stress interferes with Ca^{2+} homeostasis

Under physiological conditions, the cytosolic free Ca^{2+} concentration ($[Ca^{2+}]_i$) in mammalian cells is very low (*c.* 0.1 μM) compared with the concentration of Ca^{2+} in the extracellular fluids (1–2 mM). Such a low $[Ca^{2+}]_i$ is maintained through the concerted action of Ca^{2+}-translocases present in the mitochondrial, endoplasmic reticular, and the plasma membrane (Carafoli, 1987) and by Ca^{2+} sequestration into the nucleus as more recently proposed (Nicotera *et al.*, 1989*a*; 1990) (Fig. 2). The plasma membrane translocases (which are Ca^{2+}-ATPases in liver cells and erythrocytes, and a Na^+/Ca^{2+} exchanger in excitable tissues) balances the passive influx of Ca^{2+} from the extracellular environment, and the Ca^{2+}-translocating activities of mitochondria, endoplasmic reticulum, and perhaps also liver cell nuclei, contribute to the low $[Ca^{2+}]_i$, by active Ca^{2+} sequestration into their respective compartments.

All of these Ca^{2+} transport systems are highly sensitive to agents which affect the redox status of pyridine nucleotides and the oxidation state of protein sulphydryl groups. Consequently, a perturbation of the intracellular redox balance could result in the disruption of intracellular Ca^{2+} homeostasis. This hypothesis has received support from experiments conducted *in vivo* (Tsokos-Kuhn *et al.*, 1988), or utilising intact cells and

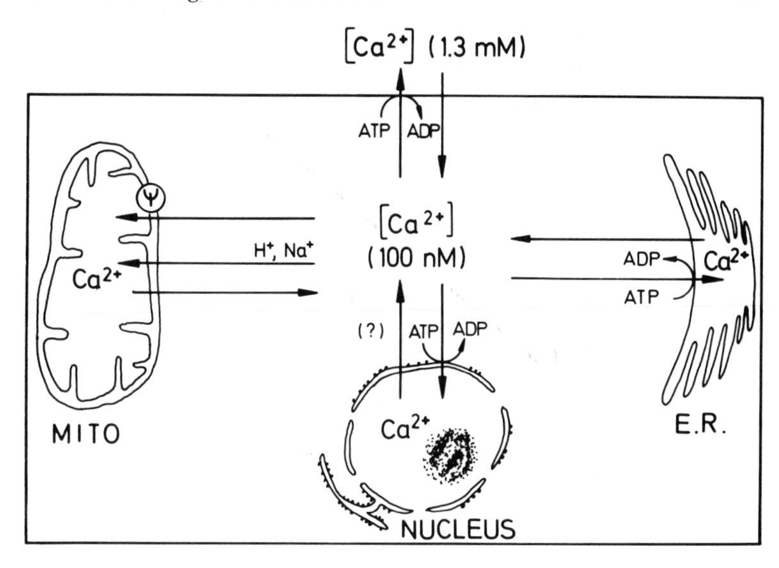

Fig. 2. Ca^{2+}-transport systems in hepatocytes.

subcellular fractions (Nicotera *et al.*, 1985; 1988; Moore, O'Brien & Orrenius, 1986; Thor *et al.*, 1985). Work from our laboratory and others has shown that during the metabolism of menadione, oxidation of pyridine nucleotides and protein thiols leads to an impairment of the Ca^{2+}-sequestering capacity of both the mitochondria and the endoplasmic reticulum. The endoplasmic reticulum is the major intracellular store of mobilisable Ca^{2+} in cells (Somlyo, Bond & Somlyo, 1985), and we have recently demonstrated that inhibition of the endoplasmic reticular Ca^{2+} translocase rapidly releases the entire endoplasmic reticular Ca^{2+} pool into the cytosol (Kass *et al.*, 1989*a*). The Ca^{2+} released by inhibition of the endoplasmic reticular Ca^{2+} translocase is only very slowly removed from the cytosol by the plasma membrane Ca^{2+} pump. In contrast to the endoplasmic reticulum, mitochondria have a low affinity for Ca^{2+}, and it is now clear that in the presence of physiological Ca^{2+} concentrations, they take up little Ca^{2+}. However, when cytosolic $[Ca^{2+}]$ is high [>1 μM], mitochondria are extremely efficient Ca^{2+} buffers, thus becoming a crucial line of defence against pathological elevations of cytosolic Ca^{2+} concentrations. During conditions of oxidative stress, the loss of both reduced pyridine nucleotides and protein thiols inhibits mitochondrial Ca^{2+} sequestration resulting in the loss of this vital defence mechanism against elevated $[Ca^{2+}]_i$, while at the same time releasing Ca^{2+} into the cytosol. If, as is often the case, the plasma membrane Ca^{2+}-

extruding system is also inhibited by thiol modification, the continuous influx of the extracellular Ca^{2+} down its concentration gradient will also contribute to the sustained elevation of cytosolic Ca^{2+} concentration observed in cells undergoing oxidative stress.

Mechanism of cell damage and cell killing during oxidative stress

A pathological elevation of $[Ca^{2+}]_i$ following inhibition of the cellular Ca^{2+} transport systems by oxidative stress, has been shown to activate several Ca^{2+}-dependent cytotoxic mechanisms. Those mechanisms include, in addition to disruption of the cytoskeletal network (see above), the pathological activation of Ca^{2+}-stimulated catabolic enzymes, such as proteases (e.g. the calpains), phospholipases and endonucleases (Fig. 3) (Orrenius *et al.*, 1989). A further notable consequence of the disruption of thiol homeostasis during exposure to cytotoxic concentrations of pro-oxidants is the impairment of hormone-stimulated phosphoinositide metabolism (Bellomo *et al.*, 1987), which will result in the isolation of the damaged cell from the hormonal control of the body.

Ca^{2+}-dependent degradative enzymes

The catabolism of phospholipids, proteins and nucleic acids involves enzymes most of which require Ca^{2+} for activity. Intracellular Ca^{2+} over-

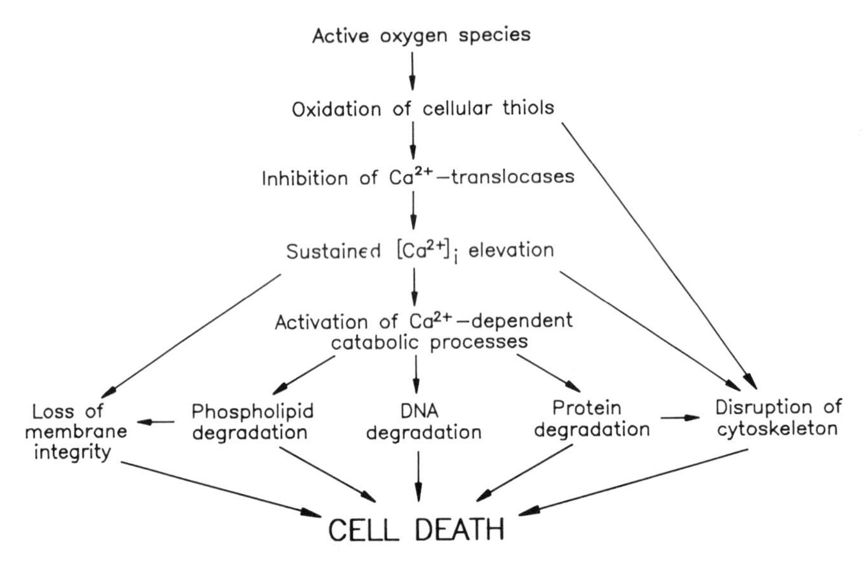

Fig. 3. Mechanisms involved in oxidative-stress-mediated cell killing.

load can result in a sustained activation of these enzymes and in the degradation of cell constituents which may ultimately lead to cell death.

Phospholipases catalyse the hydrolysis of membrane phospholipids. They are widely distributed in biological membranes and generally require Ca^{2+} for activation. A specific subset of phospholipases, collectively designated as phospholipase A_2, is Ca^{2+}- and calmodulin-dependent and thus it is susceptible to activation following an increase in cytosolic Ca^{2+} concentration. Hence, it has been suggested that a sustained increase in cytosolic Ca^{2+} can result in enhanced breakdown of membrane phospholipids and, in turn, in mitochondrial and cell damage. Although a number of studies have indicated that accelerated phospholipid turnover occurs during anoxia or toxic cell injury (Chien, Pfau & Farber, 1979; Farber & Young, 1981; Glende & Pushpendran, 1986; Nicotera *et al.*, 1989*b*) the importance of phospholipase activation in the development of cell damage remains to be established.

During the past ten years the involvement of non-lysosomal proteolysis in several cell processes has become progressively clear. Proteases which have a neutral pH optimum include the ATP- and ubiquitin-dependent proteases and the calcium-dependent proteases, or calpains. Calpains are present in virtually all mammalian cells (Murachi, 1983) and appear to be largely associated with membranes in conjunction with a specific inhibitory protein (calpastatin) (Murachi, 1983). The extra-lysosomal localisation of this proteolytic system allows the proteases to participate in several specialised cell functions, including cytoskeletal and cell membrane remodelling, receptor cleavage and turnover, enzyme activation, and modulation of cell mitosis.

Cellular targets for these enzymes include cytoskeletal elements and integral membrane proteins (Bellomo *et al.*, 1990*a*; Mellgren, 1987). Thus, the activation of Ca^{2+} proteases has been shown to cause modification of microfilaments in platelets during oxidative stress (Bellomo *et al.*, 1990*a*) and to be involved in cell degeneration during muscle dystrophy (Imahori, 1982) and in the development of ischaemic injury in nervous tissue (Manev *et al.*, 1990). Studies from our and other laboratories have suggested the involvement of Ca^{2+}-activated proteases in the toxicity of oxidants in liver (Nicotera & Orrenius, 1986; Nicotera *et al.*, 1989*b*), of venom toxins in myocardial cells (Tzeng & Chen, 1988) and of menadione in platelets (Mirabelli *et al.*, 1989). Although the substrates for protease activity during cell injury remain largely unidentified, it appears that cytoskeletal proteins may be a major target for Ca^{2+}-activated proteases during chemical toxicity.

During physiological cell killing a suicide process is activated in affected cells which is known as 'apoptosis' or programmed cell death. Several early morphological changes occur within apoptotic cells, including widespread plasma and nuclear membrane blebbing, compacting of organelles, and chromatin condensation. The most reliable and characteristic marker for this process is the activation of a Ca^{2+}-dependent endonuclease which results in the cleavage of cell chromatin into oligonucleosome-length fragments (Wyllie, 1980; Arends, Morris & Wyllie, 1990).

The results of several recent studies have shown that Ca^{2+}-overload can trigger endonuclease activation. The Ca^{2+} ionophore A23187 stimulates apoptosis in thymocytes (McConkey *et al.*, 1989), and characteristic endonuclease activity in isolated nuclei is dependent on Ca^{2+} (Jones *et al.*, 1989) and sensitive to inhibition by zinc (Arends *et al.*, 1990). In addition, Ca^{2+}-mediated endonuclease activation appears to be involved in the cytotoxicity of TCDD and tributyltin in thymocytes (McConkey *et al.*, 1988; Aw *et al.*, 1990). Although Ca^{2+}-dependent endonuclease activation has been most extensively studied in thymocytes, it appears that this process may also be important in a variety of other tissues. For example, we have identified a constitutive endonuclease in liver nuclei that is activated by submicromolar Ca^{2+} concentrations in intact nuclei incubated in the presence of ATP to stimulate Ca^{2+} uptake (Jones *et al.*, 1989). More recently, we have found that exposure of human adenocarcinoma cells to recombinant tumor necrosis factor alpha causes intracellular Ca^{2+} accumulation and endonuclease activation (Bellomo *et al.*, 1990*b*). Interestingly, in many cells the initial Ca^{2+} increase occurs in the nucleus. This suggests that selective elevation of the nuclear Ca^{2+} concentration may be sufficient to stimulate DNA fragmentation. However, although the responsible endonuclease requires Ca^{2+} for activity, its regulation appears to be more complex and to involve additional signals (Brüne *et al.*, 1990). Endonuclease activation has also been implicated in damage to macrophages caused by oxidative stress (Waring *et al.*, 1988). However, the involvement of this process in cell killing during oxidative stress remains to be elucidated.

Ca^{2+} overload may also stimulate other enzymatic processes that result in DNA damage. Elevated Ca^{2+} levels can lock topoisomerase II in a form that cleaves, but does not religate DNA, and topoisomerase II-mediated DNA fragmentation has been implicated in the cytotoxic action of some anticancer drugs (Udvardy *et al.*, 1986). DNA single-strand-breaks in cells exposed to oxidative stress can also be generated through a Ca^{2+}-dependent mechanism (Cantoni *et al.*, 1989; Dypbukt, Thor & Nicotera, 1990). Thus, further work is required not only to identify other situations in which endogenous endonuclease activation can mediate Ca^{2+}-dependent cell killing, but also to identify additional Ca^{2+}-dependent catabolic processes which may generate DNA damage.

Modulation of signal transduction pathways by oxidative stress

As opposed to the cytotoxic effects of excessive concentrations of prooxidants, recent investigations and extensive indirect evidence indicate that chemical oxidants and other agents that induce conditions of mild oxidative stress in cells can modulate cellular growth and differentiation (Sohal, Allen & Nations, 1986). In particular, oxidative stress can lead to tumor promotion and tumor progression (Cerutti, 1985). Tumor promotional activity has been demonstrated for reactive oxygen species such as H_2O_2, superoxide anion radicals, ozone, and hyperbaric oxygen, and organic peroxides including benzoylperoxide (Zimmerman & Cerutti, 1984; Slaga *et al.*, 1981). The mechanism by which agents that induce oxidative stress function as tumor promoters remains, however, unresolved.

It has been shown that conditions of oxidative stress can result in the formation of DNA strand breaks (Imlay & Linn, 1988; Dypbukt *et al.*, 1990). Interestingly, the potent skin tumor promoter, 12-*O*-tetradecanoylphorbol 13-acetate (TPA), stimulates generation of reactive oxygen species in cell types such as polymorphonuclear leukocytes (Goldstein *et al.*, 1981) and primary mouse epidermal cells (Fisher & Adams, 1985), thereby inducing DNA strand breakage (Emerit & Cerutti, 1981; Birnboim, 1982). Based on the observations that DNA strand breaks stimulate polyADP-ribosylation of chromatin proteins, Cerutti and coworkers have proposed that alteration of chromatin conformation and function through polyADP-ribosylation result in modification of expression of genes involved in the control of cell growth and differentiation, such as c-*fos* and c-*myc* (Crawford *et al.*, 1988). Interestingly, recent studies in our laboratory (Dypbukt *et al.*, 1990) and in others (Cantoni *et al.*, 1989) have indicated that intracellular Ca^{2+} chelators such as quin 2 and BAPTA prevent single-strand formation during oxidative stress, suggesting the Ca^{2+} requirement for the DNA damage by oxygen radicals. In addition, Ca^{2+} is involved in the regulation of chromatin conformation (P. Nicotera *et al.*, unpublished observations) and in conjunction with other stimuli may regulate the expression of genes such as c-*fos* and c-*myc*.

Since the expression of these genes is also modulated by protein kinase C, our laboratory has recently started to investigate the potential role of protein kinase C in oxidative stress-mediated tumor promotion.

Exposure of rat hepatocytes to low levels of oxidative stress results in a rapid, 2–3-fold increase in the specific activity of protein kinase C (Kass *et al.*, 1989*b*). Only the cytosolic protein kinase C displayed the increase in specific activity (Table 1) whereas the activity of the membrane-bound form remained unchanged. The increase in protein kinase C specific

Table 1. *Activation of hepatocyte protein kinase C during exposure to oxidative stress*

Treatment[a]	Protein kinase C activity[b]
Vehicle (DMS0)	11.2
50 μM Menadione	32.4
250 μM DMNQ[c]	27.8

Notes:
[a]Isolated rat hepatocytes were treated with redox-cycling quinones for 30 min before isolating and measuring the cytosolic protein kinase C activity.
[b]Expressed as pmol ^{32}P incorporated into histone H1/min/mg protein.
[c]DMNQ, 2,3-dimethoxy-1,4-naphthquinone.

activity is not due to an increase in protein kinase C's copy number. Rather, oxidative modification of the protein kinase C enzyme itself, most likely through oxidative-stress-induced modification of the thiol/disulphide balance of the enzyme, is responsible for the activation. The involvement of thiol residues in the activation phenomenon was confirmed when we found that partially pure protein kinase C from rat brain could be activated using low concentrations of GSSG in a glutathione redox buffer.

It is not yet established whether the oxidative modification of protein kinase C results in stimulation of substrate phosphorylation *in situ*. However, the oxidatively modified enzyme may overexpress responses to various physiological stimuli, and this may represent a potential mechanism for oxidative stress-mediated tumor promotion. Indeed, the induction of the enzyme ornithine decarboxylase (which is controlled by protein kinase C) and skin tumor promotion by TPA has been reported to be enhanced by adriamycin, a redox-cycling anthracycline antibiotic (Perchellet, Kishore & Perchellet, 1985). It should be expected that overexpression of protein kinase C-dependent responses would have a significant impact on the control of cellular growth and differentiation. Weinberg and colleagues have recently reported that overproduction of protein kinase C enzyme and overexpression of its enzyme activity are sufficient for inducing dramatic growth abnormalities, such as growth to a higher confluence density, shorter exponential doubling time, and development of foci of cells displaying anchorage-independent growth in soft agar (Housey *et al.*, 1988).

In addition to protein kinase C, other protein kinases have been shown to become activated under conditions of oxidative stress. A protein kinase which phosphorylates the α-subunit of eukaryotic initiation factor 2 is activated by low concentrations of GSSG (Ernst, Levin & London, 1978). Several laboratories have also reported that reactive oxygen species can activate the rat liver insulin receptor tyrosine kinase (Chan *et al.*, 1986; Hayes & Lockwood, 1987). More recently, induction of oxidative stress in mouse epidermal cells has been found to stimulate phosphorylation of the ribosomal protein S6 (Larsson & Cerutti, 1988) through an apparently Ca^{2+}-dependent event. Analysis of the primary structure of a number of protein kinases (serine, threonine and tyrosine kinases) has revealed a substantial sequence homology in the catalytic (kinase) domains (Parker *et al.*, 1986). Thus, a common structural feature may predispose different protein kinases to activation through alteration(s) of their thiol/disulphide status. Consequently, modulation of kinase activities involved in signal transduction and cellular metabolism could constitute a mechanism by which a non-cytotoxic state of oxidative stress is able to have profound implications on cell division and differentiation.

Summary and conclusions

Exposure of mammalian cells to oxidative stress induced by redox-active quinones and other prooxidants results in the depletion of intracellular glutathione, followed by the modification of protein thiols and the loss of cell viability. Protein thiol modification during oxidative stress is normally associated with an impairment of various cell functions, including inhibition of agonist-stimulated phosphoinositide metabolism, disruption of intracellular Ca^{2+} homeostasis, and perturbation of normal cytoskeletal organisation. The latter effect appears to be responsible for the formation of the numerous plasma membrane blebs, typically seen in cells exposed to cytotoxic concentrations of prooxidants. Following the disruption of thiol homeostasis in prooxidant-treated cells, there is an impairment of Ca^{2+} transport and a subsequent perturbation of intracellular Ca^{2+} homeostasis, resulting in a sustained increase in cytosolic Ca^{2+} concentration. This Ca^{2+} increase can cause activation of various Ca^{2+}-dependent degradative enzymes (phospholipases, proteases, endonucleases) which may contribute to cell killing. In contrast with the cytotoxic effects of excessive oxidative damage, low levels of oxidative stress can lead to the activation of a number of enzymes. In particular, the activity of protein kinase C is markedly increased by redox-cycling quinones through a thiol/disulphide exchange mechanism, and this may represent a mechanism by which prooxidants can modulate cell growth and differentiation.

References

Arends, M.J., Morris, R.G. & Wyllie, A.H. (1990). Apoptosis: The role of the endonuclease. *American Journal of Pathology* **136**, 593–608.

Aw, T.Y., Nicotera, P., Manzo, L. & Orrenius, S. (1990). Tributyltin stimulates apoptosis in rat thymocytes. *Archives of Biochemistry and Biophysics* (in press).

Bellomo, G., Mirabelli, F., Richelmi, P., Malorni, W., Iosi, F. & Orrenius, S. (1990*a*). The cytoskeleton as a target in quinone toxicity. *Free Radical Research Communications* **8**, 391–9.

Bellomo, G., Perotti, M., Taddei, F., Mirabelli, F., Finardi, G., Nicotera, P. & Orrenius, S. (1990*b*). TNF-alfa kills adenocarcinoma cells by an increase in nuclear free Ca^{2+} concentration and DNA fragmentation (submitted).

Bellomo, G., Thor, H. & Orrenius, S. (1987). Alterations in inositol phosphate production during oxidative stress in isolated hepatocytes. *Journal of Biological Chemistry* **262**, 1530–4.

Birnboim, H.C. (1982). DNA strand breakage in human leukocytes exposed to a tumor promoter, phorbol myristate acetate. *Science* **215**, 1247–9.

Brüne, B., Hartzell, P., Nicotera P. & Orrenius, S. (1990). Spermine prevents endonuclease activation and thymocyte apoptosis. *Experimental Cell Research* (in press).

Cantoni, O., Sestili, P., Cattabeni, F., Bellomo, G., Pou, S., Cohen, M. & Cerutti, P. (1989). Calcium chelator Quin 2 prevents hydrogen-peroxide-induced DNA breakage and cytotoxicity. *European Journal of Biochemistry* **182**, 209–12.

Carafoli, E. (1987). Intracellular Ca^{2+} homeostasis. *Annual Reviews of Biochemistry* **56**, 395–433.

Cerutti, P.A. (1985). Prooxidant states and tumor promotion. *Science* **227**, 375–81.

Chan, T.M., Chen, E., Tatoyan, A., Shargill, N.S., Pleta, M. & Hochstein, P. (1986). Stimulation of tyrosin-specific protein phosphorylation in the rat liver plasma membrane by oxygen radicals. *Biochemical and Biophysical Research Communications* **139**, 439–45.

Chien K.R., Pfau R.G. & Farber, J.L. (1979). Ischemic myocardic injury. Prevention by chlorpromazine of an accelerated phospholipid degradation and associated membrane dysfunction. *American Journal of Pathology* **97**, 505–30.

Crawford, D., Zbinden, I., Amstad, P. & Cerutti, P. (1988). Oxidant stress induces the proto-oncogenes c-*fos* and c-*myc* in mouse epidermal cells. *Oncogene* **3**, 27–32.

Di Monte, D., Bellomo G., Thor, H., Nicotera, P. & Orrenius, S. (1984*a*). Menadione-induced cytotoxicity is associated with protein thiol oxidation and alteration in intracellular Ca^{2+} homeostasis. *Archives of Biochemistry and Biophysics* **235**, 343–50.

Di Monte, D., Ross, D., Bellomo, G., Eklöw, L. & Orrenius, S. (1984*b*). Alterations in intracellular thiol homeostasis during the metabolism of menadione by isolated rat hepatocytes. *Archives of Biochemistry and Biophysics* **235**, 334–42.

Dypbukt, J.M., Thor, H. & Nicotera, P. (1990). Intracellular Ca^{2+} chelators prevent DNA damage and protect hepatoma 1C1C7 cells from quinone-induced cell killing. *Free Radical Research Communications* **8**, 347–54.

Emerit, I. & Cerutti, P. (1981). Tumor promoter phorbol-12-myristate-13-acetate induces chromosomal damage via indirect action. *Nature* **293**, 144–6.

Eriksson, J.E., Paatero, G.I.L., Meriluoto, J.A.O., Codd, G.A., Kass, G.E.N., Nicotera, P. & Orrenius, S. (1989). Rapid microfilament reorganization induced in isolated rat hepatocytes by microcystin-LR, a cyclic peptide toxin. *Experimental Cell Research* **185**, 86–100.

Ernst, V., Levin, D.H. & London, I.M. (1978). Inhibition of protein synthesis initiation by oxidized glutathione: activation of a protein kinase that phosphorylates the alfa subunit of eukariotic initiation factor 2. *Proceedings of the National Academy of Science USA* **75**, 4110–14.

Farber, J.L. & Young, E.E. (1981). Accelerated phospholipid degradation in anoxic rat hepatocytes. *Archives of Biochemistry and Biophysics* **221**, 312–20.

Fischer, S.M. & Adams, L.M. (1985). Suppression of tumor promoter-induced chemiluminescence in mouse epidermal cells by several inhibitors of arachidonic acid metabolism. *Cancer Research* **45**, 3130–6.

Fridovich, I. (1983). Superoxide radicals: an endogenous toxicant. *Annual Reviews of Pharmacology and Toxicology* **23**, 239–57.

Glende, E.A. Jr. & Pushpendran, K.C. (1986). Activation of phospholipase A_2 by carbon tetrachloride in isolated rat hepatocytes. *Biochemical Pharmacology* **35**, 3301–7.

Goldstein, B.D., Witz, G., Amoruso, M., Stone, D.S. & Troll, W. (1981). Stimulation of human polymorphonuclear leukocyte superoxide anion radiacal by tumor promoters. *Cancer Letters* **11**, 257–62.

Gopalakrishna, R. & Anderson, W.B. (1989). Ca^{2+}- and phospholipid-independent activation of protein kinase C by selective oxidative modification of the regulatory domain. *Proceedings of the National Academy of Science, USA* **86**, 6758–62.

Hayes, G.R. & Lockwood, D.H. (1987). Role of insulin receptor phosphorylation in the insulinomimetic effects of hydrogen-peroxide. *Proceedings of the National Academy of Science USA* **84**, 8115–19.

Housey, G.M., Johnson, M.D., Hsiao, W.-L., O'Brian, C.A., Murphy, J.P., Kirschmeier, P. & Weinstein, I.B. (1988). Overproduction of protein kinase C causes disordered growth control in rat fibroblasts. *Cell* **52**, 343–54.

Imahori, K. (1982). Calcium-dependent neutral protease: its characterization and regulation. In *Calcium and Cell Function*, vol. III, ed. W.Y. Cheung. Orlando, Fl: Academic Press.

Imlay, J.A. & Linn, S. (1988). DNA damage and oxygen radical toxicity. *Science* **240**, 1302–9.

Jewell, S.A., Bellomo, G., Thor, H., Orrenius, S. & Smith, M.T. (1982). Bleb formation in hepatocytes during drug metabolism is caused by disturbances in thiol and calcium ion homeostasis. *Science* **217**, 1257–9.

Jones, D.P., McConkey, D.J., Nicotera, P. & Orrenius, S. (1989). Calcium-activated DNA fragmentation in rat liver nuclei. *Journal of Biological Chemistry* **264**, 6398–403.

Kass, G.E.N., Duddy, S.K., Moore, G.A. & Orrenius, S. (1989*a*). Di-(*tert*-butyl)-1,4,-benzohydroquinone rapidly elevates cytosolic Ca^{2+} concentration by mobilizing the inositol 1,4,5-triphosphate-sensitive Ca^{2+} pool. *Journal of Biological Chemistry* **264**, 15192–8.

Kass, G.E.N., Duddy, S.K. & Orrenius, S. (1989*b*). Activation of protein kinase C by redox cycling quinones. *Biochemical Journal* **260**, 499–507.

Kass, G.E.N., Wright, J.M., Nicotera, P. & Orrenius, S. (1988). The mechanism of 1-methyl-4-phenyl 1,2,3,6-tetrahydropyridine toxicity. Role of intracellular calcium. *Archives of Biochemistry and Biophysics* **260**, 789–97.

Larsson, R. & Cerutti, P. (1988). Oxidants induce phosphorylation of ribosomal protein S6. *Journal of Biological Chemistry* **263**, 17452–8.

Manev, H., Costa, E., Wroblewski, J.T. & Guidotti, A. (1990). Abusive stimulation of excitatory amionoacids receptors: a strategy to limit neurotoxicity. *FASEB Journal* **4**, 2787–9.

McConkey, D.J., Hartzell, P., Duddy, S.K., Håkansson, H. & Orrenius, S. (1988). 2,3,7,8-tetrachlorobenzo-*p*-dioxin kills immature thymocytes by Ca^{2+}-mediated endonuclease activation. *Science* **242**, 256–9.

McConkey, D.J., Hartzell, P., Nicotera, P. & Orrenius, S. (1989). Calcium-activated DNA fragmentation kills immature thymocytes. *FASEB Journal* **3**, 1843–9.

Mellgren R.L. (1987). Calcium-dependent proteases: an enzyme system active at cellular membranes? *FASEB Journal* **1**, 110–15.

Mirabelli, F., Salis, A., Marinoni, V., Finardi, G., Bellomo, G., Thor, H. & Orrenius, S. (1988). Menadione-induced bleb formation in hepatocytes is associated with oxidation of thiol groups in actin. *Archives of Biochemistry and Biophysics* **264**, 261–9.

Mirabelli, F., Salis, A., Vairetti, M., Bellomo, G., Thor, H. & Orrenius, S. (1989). Cytoskeletal alterations in human platelets exposed to oxidative stress are mediated by oxidative and Ca^{2+}-dependent mechanisms. *Archives of Biochemistry and Biophysics* **270**, 478–88.

Moore, G.A., O'Brien, P.J. & Orrenius, S. (1986). Menadione (2-methyl-1,4-naphthoquinone)-induced Ca^{2+} release from rat-liver mitochondria is caused by NAD(P)H oxidation. *Xenobiotica* **16**, 873–82.

Muehlematter, D., Larsson, R. & Cerutti, P. (1988). Active oxygen-induced DNA strand breakage and poly-ADP-ribosylation in promotable and non-promotable JB6 mouse epidermal cells. *Carcinogenesis* **9**, 239–45.

Murachi, T. (1983). Intracellular Ca^{2+} proteases and its inhibitor protein: calpain and calpastatin. In *Calcium and Cell Function* vol. IV, ed. W.Y. Cheung, pp. 376–410. Orlando, Fl: Academic Press.

Nicotera, P., Hartzell, P., Davis, G. & Orrenius, S. (1986). The formation of plasma membrane blebs in hepatocytes exposed to agents that increase cytosolic Ca^{2+} is mediated by the activation of a non-lysosomal proteolytic system. *FEBS Letters* **209**, 139–44.

Nicotera, P., McConkey, D.J., Jones, D.P. & Orrenius, S. (1989*a*). ATP-stimulates Ca^{2+} uptake and increases the free Ca^{2+} concentration in isolated liver nuclei. *Proceedings of the National Academy of Science USA* **86**, 453–7.

Nicotera, P., McConkey, D., Svensson, S.Å., Bellomo, G. & Orrenius, S. (1988). Correlation between cytosolic Ca^{2+} concentration and cytotoxicity in hepatocytes exposed to oxidative stress. *Toxicology* **52**, 55–63.

Nicotera, P., Moore, M., Mirabelli, F., Bellomo, G. & Orrenius, S. (1985). Inhibition of hepatocyte plasma membrane Ca^{2+}-ATPase activity by menadione metabolism and its restoration by thiols. *FEBS Letters* **181**, 149–53.

Nicotera, P. & Orrenius, S. (1986). Role of thiols in protection against biological reactive intermediates. In *Biological Reactive Intermediates III*, ed. J.J. Kocsis, D.J. Jollow, C.M. Witmer, J.O. Nelson & R. Snyder, pp. 41–51. New York: Plenum.

Nicotera, P., Orrenius, S., Nilsson, T. & Berggren, P.O. (1990). An inositol 1,4,5-trisphosfate-sensitive Ca^{2+} pool in liver nuclei. *Proceedings of the National Academy of Science USA* **87**, 6858–62.

Nicotera, P., Rundgren, M., Porubek, D., Cotgreave, J., Moldéus, P., Orrenius, S. & Nelson, S.D. (1989*b*). On the role of Ca^{2+} in the toxicity of alkylating and oxidizing quinoneimines in isolated hepatocytes. *Chemical Research in Toxicology* **2**, 46–50.

Orrenius, S., McConkey, D.J., Bellomo, G. & Nicotera, P. (1989). Role of Ca^{2+} in toxic cell killing. *Trends in Pharmacological Sciences* **10**, 281–5.

Orrenius, S. & Moldéus, P. (1984). The multiple role of glutathione in drug metabolism. *Trends in Pharmacological Sciences* **5**, 432–5.

Orrenius, S., Rossi, L., Eklöw-Låstbom, L. & Thor, H. (1986). In *Free Radicals in Liver Injury*, ed. G. Poli & K.H. Cheesman, pp. 99–105. Oxford: IRL Press Ltd.

Parker, P.J., Coussens, L., Totty, N., Rhee, L., Young, S., Chen, E., Stabel, S., Waterfield, M.D. & Ullrich A. (1986). The complete primary structure of protein kinase C – the major phorbol ester receptor. *Science* **233**, 853–9.

Perchellet, J.P., Kishore, G.S. & Perchellet, E.M. (1985). Enhancement by adriamycin of the effect of 12-o-tetradecanoylphorbol-13-acetate on mouse epidermal glutathione peroxidase activity, ornithine decarboxylase induction and skin tumor promotion. *Cancer Letters* **29**, 127–37.

Prentki, M., Chaponnier, C., Jeanrenaud, B. & Gabbiani, G. (1979). Actin microfilaments, cell shape, and secretory processes in isolated rat hepatocytes: effect of phalloidin and cytochalasin D. *Journal of Cell Biology* **81**, 592–607.

Sies, H. (ed.) (1985). Oxidative stress: introductory remarks. In *Oxidative Stress* pp. 1–8. London: Academic Press.

Sies, H., Brigelius, R. & Graf, P. (1987). Hormones, glutathione status and protein-S-thiolation. In *Advances in Enzyme Regulation*, vol. 26, ed. G. Weber, pp. 175–89. Oxford: Pergamon Press.

Slaga, T.J., Klein-Szanto, A.J.P., Triplett, L.L., Yotti, L.P. & Trosko, J.E. (1981). SKin-tumor promoting activity of benzoyl peroxide, a widely used free radical-generating compound. *Science* **213**, 1023–4.

Sohal, R.S., Allen, R.G. & Nations, C. (1986). Oxygen free radicals play a role in cellular differentiation: an hypothesis. *Journal of Free Radicals in Biology and Medicine* **2**, 175–81.

Somlyo, A.P., Bond, M. & Somlyo, A.V. (1985). Calcium content of mitochondria and endoplasmic reticulum in liver frozen rapidly *in vivo*. *Nature* **314**, 622–5.

Smith, M.T., Evans, C.G., Thor, H. & Orrenius, S. (1985). Quinone-induced oxidative injury to cells and tissues. In *Oxidative Stress*, ed. H. Sies, pp. 91–113. London: Academic Press.

Thor, H., Hartzell, P., Svensson, S.-Å., Orrenius, S., Mirabelli, F., Marinoni, V. & Bellomo, G. (1985). On the role of thiol groups in the inhibition of liver microsomal Ca^{2+} sequestration by toxic agents. *Biochemical Pharmacology* **34**, 3717–23.

Thor, H., Smith, M.T., Hartzell, P., Bellomo, G., Jewell, S.A. & Orrenius, S. (1982). The metabolism of menadione by isolated hepatocytes. A study of the implications of oxidative stress in intact cells. *Journal of Biological Chemistry* **257**, 12419–25.

Tsokos-Kuhn, J.O., Sith, C.V., Hughes, H. & Mitchell, J.R. (1988). Liver membrane calcium transport in diquat-induced oxidative stress *in vivo*. *Molecular Pharmacology* **34**, 209–14.

Tzeng, W.F. & Chen, Y.H. (1988). Suppression of snake-venom cardiomyocyte degeneration by blockage of Ca^{2+} influx or inhibition of non-lysosomal proteinases. *Biochemical Journal* **256**, 89–95.

Udvardy, A., Schedl, P., Sander, M. & Hsieh, T. (1986).

Topoisomerase II cleavage in chromatin. *Journal of Molecular Biology* **191**, 231–46.
Waring, P., Eichner, R.D., Müllbacher, A. & Sjaarda, A. (1988). Gliotoxin induces apoptosis in macrophages unrelated to its antiphagocytic properties. *Journal of Biological Chemistry* **263**, 18493–9.
Weeds, A. (1982). Actin-binding proteins regulators of cell architecture and motility. *Nature* **296**, 811–16.
Wyllie, A.H. (1980). Glucocorticoid-induced thymocyte apoptosis is associated with endogenous endonuclease activation. *Nature* **280**, 555–6.
Zimmerman, R. & Cerutti, P. (1984). Active oxygen acts as a promotor of transformation in mouse embryo C3H/10 T $\frac{1}{2}$/C18 fibroblasts. *Proceedings of the National Academy of Science USA* **81**, 2085–7.

STEVEN W. EDWARDS

Regulation of neutrophil oxidant production

Polymorphonuclear leukocytes (neutrophils) provide the first line of defence to protect the host against most bacterial and many fungal pathogens, and hence possess an array of specialised cytotoxic processes and associated pathways in order to perform this important function during phagocytosis. In the 1930s the importance of O_2 was recognised when it was discovered (Baldridge & Gerard, 1933) that a 'respiratory burst' accompanied phagocytosis, but this was mistakenly believed to be due to increased mitochondrial respiration necessary to supply the extra energy required for phagocytosis. The unusual nature of the respiratory burst was not appreciated until later (Sbarra & Karnovsky, 1959) when it was found to be uninhibited by cyanide and hence not associated with mitochondrial respiration. Later, it was proposed that H_2O_2 was generated during phagocytosis (Iyer, Islam & Quastel, 1961) and that $O_2^{\overline{\cdot}}$ was the primary product of O_2 reduction (Babior, Kipnes & Curnutte, 1973). It was therefore suggested that the respiratory burst was required for the generation of oxygen metabolites which were instrumental in pathogen killing (Selvaraj & Sbarra, 1966). The link between the products of the burst and microbial killing was confirmed when it was discovered that phagocytes from patients with chronic granulomatous disease (CGD, formerly Fatal Granulomatous Disease of Childhood) who are predisposed to life-threatening infections had an impaired ability of their phagocytes to mount a respiratory burst (Holmes, Page & Good, 1967). The search for the molecular defects responsible for this condition (or group of related conditions) has proved invaluable in identifying the molecular components required for oxidant generation (Segal, 1989*a*,*b*), and also in elucidating the complex processes by which these become assembled and activated during phagocytosis.

In addition to the products of the respiratory burst, neutrophils also possess a variety of other cytotoxic components which undoubtedly play a role in microbial killing (Spitznagel, 1984; Lehrer *et al.*, 1988). These are located within subcellular granules (membrane-bound organelles) which fuse with the phagocytic vesicle containing the engulfed pathogen. The contents of these granules then discharge into the phagolysosome (degranulation) where they may interact with the products of the respiratory burst. These granule enzymes include: myeloperoxidase, a haemoprotein which reacts with H_2O_2 (and $O_2^{\overline{\cdot}}$) to oxidise halides to hypohalous acids and related compounds such as chloramines (Klebanoff, 1968; Klebanoff & Clark, 1978); proteases such as elastase and cathepsin G (Boxer & Smolen, 1988); permeability-inducing factors such as BPI (bacterial permeability inducing protein, Elsbach & Weiss, 1983); defensins, which are small cationic peptides (Ganz *et al.*, 1985); and lysozyme.

Much controversy exists concerning the precise role played by these different cytotoxic processes (oxidants and granule enzymes) in microbial killing, as the biochemical reactions which occur within the phagolysosome and which result in target death are yet to be defined. Undoubtedly neutrophils possess a degree of 'overkill' in order to ensure maximal host protection against a range of pathogen targets which may have varying degrees of protection against any **one** cytotoxic system. This is demonstrated by the fact that different target organisms show considerable variation in their susceptibility to these different cytotoxic systems *in vitro*, and neutrophils from CGD patients can efficiently kill **some** pathogens (presumably via their granule enzymes) whereas others cannot be killed (Tauber *et al.*, 1983).

Reactive oxidant generation

The respiratory burst

It is now recognised that the respiratory burst enzyme is a plasma-membrane-bound NADPH oxidase which is normally dormant in resting cells (Babior, 1978; 1984*a*,*b*; Rossi, 1986; Curnutte, 1988). Upon phagocytosis (or other forms of cell stimulation), this oxidase becomes activated by a series of complex intracellular processes (Fig. 1) and reduces O_2 in a one-electron step to form $O_2^{\overline{\cdot}}$ which is the primary product of the burst. NADPH is believed to be the substrate *in vivo* and this is provided by the increased activity of the hexose monophosphate shunt which accompanies cell stimulation. The NADPH oxidase is, in fact, a multi-component enzyme complex which constitutes an electron transfer chain from NADPH and terminates in O_2 reduction. Thus, the overall reaction catalysed by the oxidase is:

$$NADPH + 2O_2 \rightarrow 2O_2^{\overline{\cdot}} + NADP^+ + H^+ \quad (1)$$

The $O_2^{\cdot-}$ generated may then dismutate either enzymically or spontaneously:

$$2O_2^{\cdot-} + 2H^+ \rightarrow O_2 + H_2O_2 \quad (2)$$

Further reactions of O_2^- and H_2O_2 are possible and in the presence of transition metal salts such as iron and copper salts, a complex series of redox reaction may occur, summarised as a metal-catalysed Haber–Weiss type reaction (Halliwell & Gutteridge, 1984; 1985):

$$O_2^{\cdot-} + H_2O_2 \rightarrow {}^{\cdot}OH + {}^{-}OH + O_2 \quad (3)$$

The $^{\cdot}OH$ which may be formed in this reaction is one of the most reactive species known in biological systems. Some evidence from EPR spectroscopy using spin trapped adducts suggests that $^{\cdot}OH$ may indeed be formed by activated neutrophils (Bannister & Bannister, 1985; Britigan, Cohen & Rosen, 1987) although caution must be exercised in interpreting such data because $O_2^{\cdot-}$ generated products may decay to form adducts resembling those generated by $^{\cdot}OH$ directly. For example, 5,5-dimethyl-1-pyrroline-1-oxide(DMPO) can react with $O_2^{\cdot-}$ to form DMPO—OOH, and with $^{\cdot}OH$ to form DMPO—OH; formation of the latter adduct in phagocytosing neutrophils is taken as evidence for $^{\cdot}OH$ formation. However, two facts must be considered. Firstly, DMPO—OH formation must be catalase-sensitive as $^{\cdot}OH$ formation requires H_2O_2, as in reaction 3, above. Secondly, DMPO—OH may be formed from the decomposition of DMPO—OOH. It has recently been shown (Britigan *et al.*, 1989) that catalase-sensitive (DMPO—OOH independent) DMPO—OH formation in neutrophils can be detected and that $^{\cdot}OH$ formation may be regulated by the extent of specific- and azurophilic-degranulation: lactoferrin (from specific granules) can reduce $^{\cdot}OH$ formation by chelating iron, whereas myeloperoxidase (from azurophilic granules) can reduce $^{\cdot}OH$ formation by utilising H_2O_2 (Winterbourn, 1986; Cohen *et al.*, 1988; Britigan *et al.*, 1989). However, it must be stressed that $^{\cdot}OH$ formation in these experiments is only detected when neutrophil suspensions are supplemented with exogenous iron in the form of iron-diethylenetriaminepenta-acetic acid (Britigan *et al.*, 1989; Pou *et al.*, 1989).

Myeloperoxidase, normally present in the azurophilic granules (Klebanoff & Clark, 1978) but which discharges to the site of oxidant production, greatly potentiates the toxicity of H_2O_2 by virtue of its ability to oxidise halides such as Cl^-, e.g. in the chlorination reaction:

$$H_2O_2 + Cl^- \rightarrow OCl^- + H_2O \quad (4)$$

The HOCl produced is a powerful oxidant and highly bactericidal;

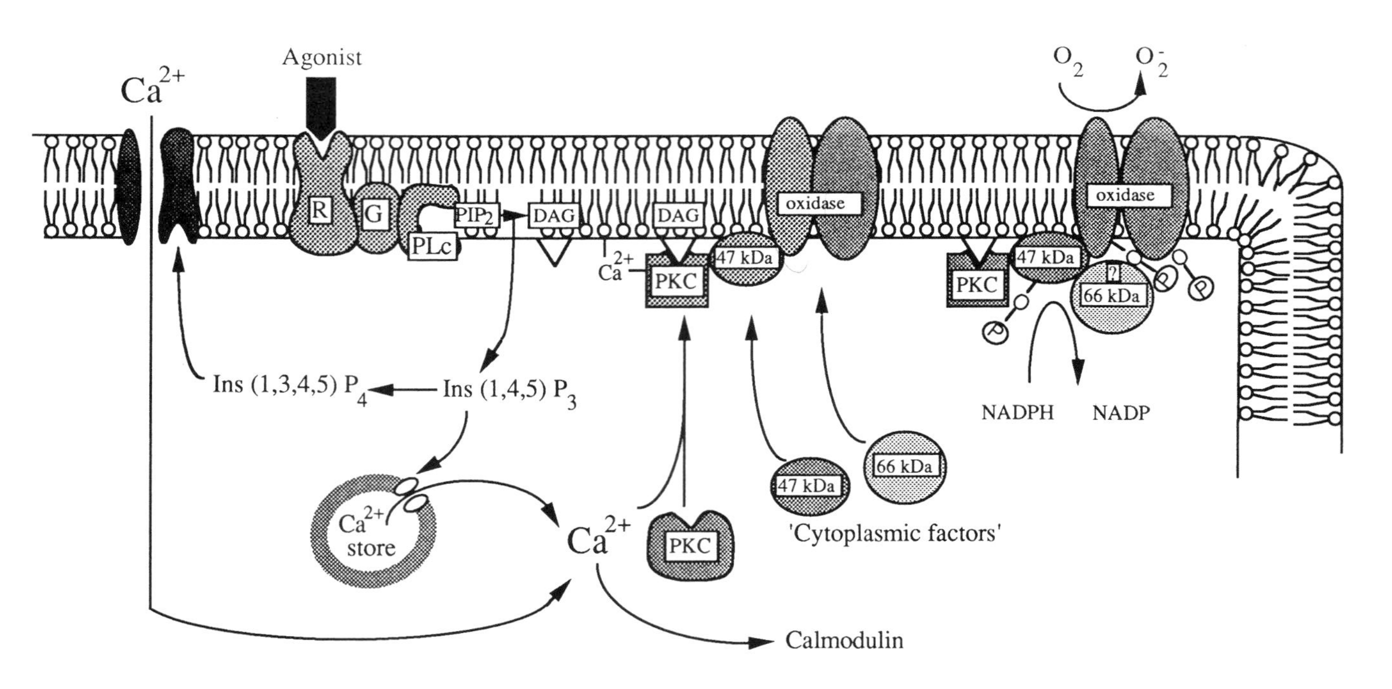

Agonist
Ca^{2+}
R
G
PLc
PIP_2
DAG
DAG
Ca^{2+}
PKC
47 kDa
oxidase
O_2
O_2^-
oxidase
PKC
47 kDa
?
66 kDa
P
P
P
NADPH
NADP
Ins (1,3,4,5) P_4
Ins (1,4,5) P_3
Ca^{2+} store
Ca^{2+}
PKC
47 kDa
66 kDa
'Cytoplasmic factors'
Calmodulin

oxidants produced from this such as chloramines are longer lived and those formed from lipophilic amines may permeate membranes:

$$OCl^- + R{-}NH_2 \rightarrow R{-}NHCl + OH^- \quad (5)$$

Myeloperoxidase may also react with H_2O_2 and $O_2^{\bar{\cdot}}$ to form compounds 'I, II and III', which are spectrally-identifiable redox states of myeloperoxidase, depending upon the concentrations of oxidants and halides in its micro-environment (Odajima & Yamazaki, 1972; Winterbourne, Garcia & Segal, 1985; Edwards & Lloyd, 1986; Kettle & Winterbourn, 1988), and complex redox reactions involving this enzyme may occur in order to prevent catalytically-inactive forms accumulating.

Properties of the NADPH oxidase

The characterisation of the components of this complex have been severely hampered by problems of its solubilisation and instability, particularly in the presence of salts. Despite these methodological problems a clearer understanding of its molecular organisation is now developing, aided by the analysis of patients with CGD, a condition in which either key components of the oxidase are absent or else fail to become activated (Segal, 1988; 1989*a*,*b*). The picture which is emerging is that the oxidase

---

Fig. 1. Activation and assembly of the NADPH oxidase during receptor-mediated neutrophil stimulation. The molecular events following binding of an agonist to its receptor (R) leading to activation of the $O_2^{\bar{\cdot}}$-generating NADPH oxidase are schematically represented. After receptor occupancy, phospholipase C (Plc) is activated (via G-protein (G) coupling) to release diacylglycerol (DAG) and inositol 1,4,5-trisphosphate (Ins (1,4,5) P_3). The latter compound moves into the cytosol where it may release Ca^{2+} from its intracellular store or else become phosphorylated to yield inositol 1,3,4,5-tetrakisphosphate (Ins (1,3,4,5) P_3) which may open the Ca^{2+} 'gate' and lead to a sustained elevation of cytosolic Ca^{2+} levels. The increased Ca^{2+} levels may 'activate' calmodulin and other Ca^{2+}-dependent processes. Protein kinase C (PKC) translocates from the cytosol to the plasma membrane and forms a quaternary complex with DAG and Ca^{2+}. Cytoplasmic factors (e.g. 47 and 66 kDa proteins) translocate to the plasma membrane and integrate with oxidase components (including cytochrome *b*) to form an 'active' $O_2^{\bar{\cdot}}$-generating complex. Phosphorylation of the cytochrome subunits and cytosolic factors takes place, that of the latter may occur prior to their translocation. Other pathways of activation of the oxidase via phospholipases A_2 and D also exist but these are not as clearly defined. For details see text.

is an extraordinarily complicated structure which requires a number of membrane and cytosolic components functioning coordinately, and which is activated by multiple and complex routes (Fig. 1). Because reactive oxidants generated directly or indirectly by this oxidase are potentially hazardous to host tissues, perhaps such complex processes to specifically switch the oxidase on and off are necessary to reduce the possibility of tissue damage resulting from inappropriate or prolonged activation.

Components of the oxidase

Cytochrome *b*

The first molecular component of the NADPH oxidase to be identified was an unusual cytochrome *b*. In 1978 it was shown that a *b*-type cytochrome, previously noted in horse granulocytes (Hattori, 1961) became associated with the phagolysosomes (Segal & Jones, 1978) and was also absent in patients with the X-linked form of CGD (Segal *et al.*, 1978). Throughout the late 1970s and 1980s this cytochrome *b* was purified, partially characterised and its gene cloned. Its mid-point redox potential of −245 mV (Cross *et al.*, 1981) is close to that of the $O_2/O_2^{\bar{\cdot}}$ couple and hence low enough to reduce directly O_2 to $O_2^{\bar{\cdot}}$ (Wood, 1987), and it has since been discovered in a number of other 'professional phagocytes' such as monocytes, macrophages and eosinophils (Segal *et al.*, 1981). It has an absorption maximum at 558 nm in reduced–oxidised difference spectra and hence is referred to as cytochrome b_{558}, b_{-245} or *b*-245. In unstimulated neutrophils the cytochrome is present in the plasma membrane and in membranes of specific granules, but upon activation these granules fuse with the plasma membrane to translocate the cytochrome from its sub-cellular location (Borregaard *et al.*, 1983*a*; Garcia & Segal, 1984).

There has been much debate as to the role played by this cytochrome in oxygen reduction and in particular whether it is the terminal oxidase of the electron transfer chain. Certainly the available evidence supports this function: it binds CO (Cross *et al.*, 1981); when reduced its half time of oxidation by O_2 is 4.7 ms (Cross *et al.*, 1982*a*); it contributes to photochemical action spectra for the relief of CO-inhibited respiration (Edwards & Lloyd, 1987). Whilst NADPH only reduces this cytochrome very slowly under anaerobic conditions (Cross *et al.*, 1982*a*), steady-state levels of reduction in the presence of O_2 closely correlate with the kinetics of $O_2^{\bar{\cdot}}$ generation (Cross, Parkinson & Jones, 1985). Hence the evidence in favour of its role as the terminal O_2-reducing redox component of the oxidase is extremely strong.

The cytochrome consists of two components, namely a 76–92 kDa glycoprotein (the β subunit) and a smaller 23 kDa (α) subunit (Harper *et al.*, 1984; 1985) and this latter component probably binds the haem moiety (Nugent, Gratzer & Segal, 1989; Yamaguchi *et al.*, 1989). The genes for both subunits of this cytochrome have been cloned (Royer-Pokora *et al.*, 1986; Parkos *et al.*, 1988), that of the β subunit by elegant reverse genetics, and these possess little sequence homology with other cytochrome oxidases and *b*-cytochromes. Transcription of the β subunit is largely restricted to cells of the myeloid lineages (Newburger *et al.*, 1988; Cassatella *et al.*, 1989), and the α subunit is transcribed in other non-phagocytic cells which do not express a spectrally-identifiable component (Parkos *et al.*, 1988). Recently it has been shown that some transformed B-lymphocyte cell lines also express this cytochrome (Maly *et al.*, 1988; 1989; Hancock, Maly & Jones, 1989) and the cytochrome is expressed on the cell surface of peripheral blood B-lymphocytes (Kobayashi *et al.*, 1990). In almost all cases of X-linked CGD a spectrum for this cytochrome is absent, but present in the autosomal recessive form of this disease (Segal *et al.*, 1983; Bohler *et al.*, 1986; Ohno *et al.*, 1986; Segal, 1987). In X-linked, cytochrome *b* negative CGD (X-CGD or X^--CGD) both the α and β subunits are absent even though the genetic defect is restricted to the heavy chain (Dinauer *et al.*, 1987; Parkos *et al.*, 1987; Segal, 1987): hence, the expression of the heavy chain is necessary for the expression/translation/stabilisation of the light chain. A small number of X-linked CGD patients have been identified which do show a spectrum for cytochrome *b* (Borregaard *et al.*, 1983*b*; Bohler *et al.*, 1986; Okamura *et al.*, 1988) and these have been termed X^+-CGD: in a family of X^+-CGD a point mutation conferring a Pro→His mutation has been discovered and this is proposed to alter the ability of the assembled cytochrome to function correctly (Dinauer *et al.*, 1989). Both subunits of this cytochrome become phosphorylated during neutrophil activation, but this occurs after activation of O_2^- generation and hence this process in itself cannot be rate-limiting for oxidase activation (Garcia & Segal, 1988).

Flavoproteins

The participation of flavoproteins as components of the NADPH oxidase have long been predicted and experimental evidence to date favours this idea with FAD being the likely co-factor. Data which support a role for flavoproteins in oxidase activity are: solubilised oxidase preparations are inhibited by flavin analogues such as 5-carbadeaza-FAD (Light *et al.*, 1981) or the flavoprotein inhibitor DPI (Cross & Jones, 1986); oxidase-rich solubilised membrane preparations possess flavins (Cross *et al.*,

1982*b*; Gabig, 1983; Borregaard & Tauber, 1984; Rossi, 1986); a flavin semiquinone EPR spectrum is observed in membranes of activated neutrophils (Kakinuma *et al.*, 1986); flavin levels in the membranes of neutrophils from many patients with autosomal recessive CGD are reduced to about half that found in normal controls (Cross *et al.*, 1982*b*; Gabig & Lefker, 1984; Bohler *et al.*, 1986; Ohno *et al.*, 1986).

Several groups have reported that a polypeptide of 65–66 kDa, which co-purifies with oxidase activity, contains flavin (Doussiere & Vignais, 1985; Markert, Glass & Babior, 1985; Kakinuma, Fukuhara & Kaneda, 1987). A 45 kDa polypeptide is labelled by the flavoprotein- and oxidase-inhibitor, DPI (Cross & Jones, 1986) and NADPH appears to compete for the DPI-binding site. This component has now been purified and has been shown to be an FAD-binding protein: polyclonal antibodies raised to it co-precipitate both this 45 kDa flavoprotein and the small (α) subunit of cytochrome b_{-245} (Yea, Cross & Jones, 1990). Hence it is proposed that this flavoprotein is the redox carrier transferring electrons from NADPH to cytochrome *b* (Ellis, Cross & Jones, 1989):

$$\text{NADPH} \rightarrow \text{Flavoprotein} \rightarrow \text{cytochrome } b_{-245} \rightarrow O_2$$

Other components

Quinones (Crawford & Schneider, 1982; Cunningham *et al.*, 1982; Gabig & Lefker, 1985) have been implicated as redox components of the oxidase, but evidence in favour of their involvement is weak. Furthermore, oxidase preparations containing other putative components have been described (e.g. Babior, 1988), but at present no molecular probes are available and hence their functional role cannot be assigned. However, it is established that a number of cytosolic components (described below) translocate to membrane fractions during oxidase activation (including 47 and 66 kDa proteins), and their functional involvement is unquestioned.

Activation of the oxidase

Receptors

The link between the external environment or pathogenic target of the neutrophil, and the activation of its varied functions (chemotaxis, adherence, phagocytosis, oxidase activation and degranulation) is provided by the plasma membrane receptors. Hence, these receptors are coupled to complex intracellular signal transduction systems to provide the appropriate cellular response upon occupancy. These receptors can be rapidly up-regulated in response to external signals either by changes in the numbers expressed on the plasma membrane (via mobilisation of pre-formed subcellular pools) or by covalent modifications (e.g. phosphorylations) which alter their affinities: conversely, receptor expression may be rapidly down-regulated by covalent modification, internalisation or shedding. Thus, neutrophil function *in vivo* will be regulated by the levels of expression of these receptors and how they are coupled to their respective signal transduction systems. Specific receptors have been described for all of the major neutrophil activating agonists such as fMet–Leu–Phe, C5a, LTB_4, PAF, neutrophil activating peptide (NAP or IL-8) and several cytokines (such as G-CSF, GM-CSF and γ-interferon). The complement and IgG receptors which are involved in adherence and opsono-phagocytosis are best characterised at the molecular level.

1. IgG Fc receptors

Human leukocytes possess three types of surface receptors specific for the Fc domains of immunoglobulin G, which are distinguished by differences in their physiochemical properties and which are recognised by unique sets of monoclonal antibodies (Anderson & Looney, 1986).

FcRI (CD64) is a 72 kDa glycoprotein found in monocytes/macrophages and U937 cells, and binds monomeric IgGl and IgG3 with high affinity (Anderson & Abraham, 1980; Anderson & Looney, 1986). It is only expressed in neutrophils after γ-interferon exposure (Perussia *et al.*, 1983; Petroni, Shen & Guyre, 1988).

FcRII (CD32) is a 40 kDa component of neutrophils, monocytes, B cells, eosinophils and platelets which binds monomeric IgG with low affinity but binds well to aggregated IgG (Jones, Looney & Anderson, 1985; Looney, Abraham & Anderson, 1986). There are about 10K–20K receptors per neutrophil and this receptor is responsible for activation of phagocytosis and oxidant generation (Huizinga *et al.*, 1989).

FcRIII (CD16) appears as a broad band of 50–70 kDa in SDS-PAGE and is expressed in neutrophils, eosinophils, macrophages and NK cells. It binds IgG with low affinity and is responsible for IgG binding, but its occupancy does not result in phagocytosis or oxidase activation (Huizinga *et al.*, 1989). It is linked to the plasma membrane via an N-terminal glycosylphosphatidyl inositol group (Huizinga *et al.*, 1988; Simmons & Seed, 1988; Selvaraj *et al.*, 1988), and may be shed from the plasma membrane during activation via cleavage of this linkage. There are approx. 100K–200K receptors per neutrophil.

2. Adhesion receptors
A family of related surface glycoproteins (LFA-1) is required for adherence-related leukocyte functions such as aggregation, adherence to endothelial cells, chemotaxis and phagocytosis (Anderson & Springer, 1987; Arnaout, 1990). This family (CD11/CD18) comprises three heterodimeric subunits with distinct heavy (or α) chains, but sharing a common light (β) subunit (Kurzinger & Springer, 1982; Sanchez-Madrid *et al.*, 1983; Shaw, 1987). These are: LFA-1 (CD11a), a 95 kDa polypeptide; Mac-1 (or Mo-1, CD11b) of 165 kDa; pl50,95 (CD11c). The common β-chain (CD18) is a 95 kDa polypeptide. Patients with defects in these glycoproteins (Arnaout *et al.*, 1984; Anderson *et al.*, 1985; Anderson & Springer, 1987; Todd & Freyer, 1988) have impaired neutrophil function *in vivo* resulting in increased susceptibility to infections.
In neutrophils, Mac-1 (CD11b/CD18) is the major adhesion glycoprotein of this family and is the receptor for C3bi (Beller, Springer & Schreiber, 1982; Sanchez-Madrid *et al.*, 1983; Wright *et al.*, 1983). There are normally about 10K–20K receptors per cell but this number can be rapidly upregulated during activation (Berger *et al.*, 1984; Todd *et al.*, 1984; Ross & Medof, 1985; Miller *et al.*, 1987) or priming (Arnaout *et al.*, 1986). This increased surface expression occurs rapidly (within 10–15 min), is independent of *de novo* protein biosynthesis and results from the translocation of pre-formed receptors from a sub-cellular pool to the plasma membrane (O'Shea *et al.*, 1984). These sub-cellular pools are present on specific (Todd *et al.*, 1984; Stevenson, Nauseef & Clark, 1987) or gelatinase-containing granules (Petrequin *et al.*, 1987; Jones *et al.*, 1988), but this rapid upregulation is neither necessary nor sufficient to explain increased neutrophil adhesion to endothelial cells during activation (Vedder & Harlan, 1988; Philips *et al.*, 1988).

Signal transduction

The molecular processes which couple receptor occupancy to activation of neutrophil functions (such as oxidase activation) closely resemble those which have been identified and characterised in other excitable cell

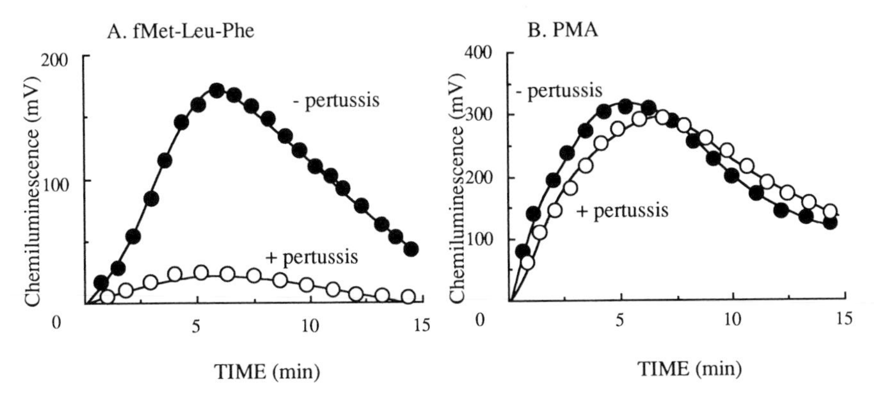

Fig. 2. Effect of pertussis toxin on neutrophil oxidant generation. Human neutrophils (10^6/ml, suspended in RPMI 1640 medium) were incubated at 37 °C for 3 h in the presence (○) and absence (●) of pertussis toxin (0.5 μg/ml). After incubation, 10 μM luminol was added to each suspension and chemiluminescence measured after the addition of 1 μM fMet–Leu–Phe (A) or 0.1 μg/ml PMA (B), final concentrations. Experiment of F. Watson & J. Robinson.

types. Signal transduction following neutrophil activation by the chemotactic peptide fMet–Leu–Phe has been most extensively studied as this agent can induce a number of neutrophil responses (such as chemotaxis, shape changes, adherence, oxidase activation, degranulation and priming) depending upon the concentration used. The series of events following binding of an agonist (such as fMet–Leu–Phe) to its receptor are outlined schematically in Fig. 1, but as we shall see later, alternative signal transduction systems must also exist although these, at present, are not clearly defined.

The fact that pertussis toxin inactivates GTP-binding proteins of the G_1 type by ADP-ribosylation of the α subunit (Okajima, Katada & Ui, 1985; Neer & Clapham, 1988) can be exploited to confer pertussis toxin sensitivity as representing G-protein mediated signal transduction. Similarly, aluminium fluoride activates G-proteins in the absence of GTP (Gabig *et al.*, 1987), and the non-hydrolysable GTP analogue GTPγS can stimulate G-protein mediated pathways in permealised cells (Nasmith, Mills & Grinstein, 1989). Hence, these approaches have been adopted to show that many receptor-mediated activation events in neutrophils, including oxidase activation, are coupled through G-proteins (Fig. 2). The G-protein then activates a phosphatidylinositol-4,5-bisphosphate to generate two second messages: inositol-1,4,5-trisphosphate (IP_3) which is released into the cytosol where it triggers the release of Ca^{2+} from intracellular stores leading to a transient rise in Ca^{2+} levels (Berridge & Irvine, 1984; 1989), and diacylglycerol (DAG) which remains in the membrane and forms a quaternary complex with Ca^{2+} and protein kinase C, which translocates from the cytosol (Tauber, 1987), probably as a result of the increased Ca^{2+} transient.

Roles of Ca^{2+} and protein kinase C in neutrophil activation

Evidence favouring the role of intracellular Ca^{2+} transients in receptor-mediated neutrophil activation leading to oxidase activation is strong. Experiments in which intracellular Ca^{2+} levels are monitored by photoproteins (Hallett & Campbell, 1983) or fluorescent indicators (Pozzan *et al.*, 1982) have shown that transients in the intracellular levels of this cation precede oxidase activation (Hallett, Edwards & Campbell, 1987; Al-Mohanna & Hallett, 1988). Furthermore, in neutrophils in which intracellular levels of Ca^{2+} are buffered by Quin-2 loading, $O_2^{\cdot-}$ secretion is inhibited (Fig. 3). Extracellular Ca^{2+} is required to sustain the levels of signal after mobilisation of the intracellular store and inositol-1,3,4,5-tetrakisphosphate may open the Ca^{2+} channel (Sadler & Badwey, 1988).

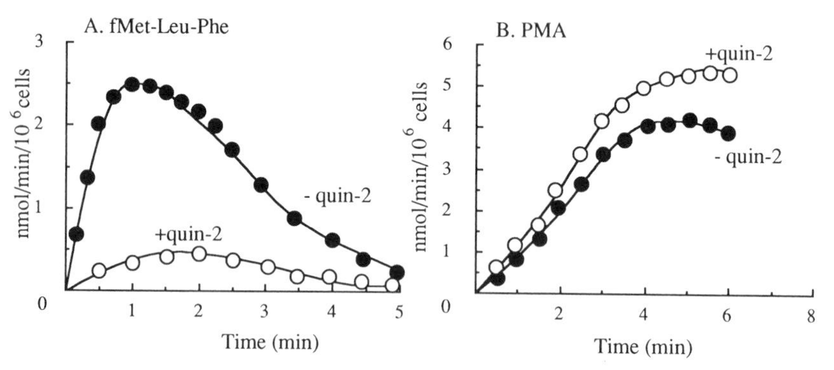

Fig. 3. Dependence of $O_2^{\cdot-}$ secretion upon intracellular Ca^{2+}. Neutrophils (10^8/ml) were incubated for 10 min in the presence (○) and absence (●) of 500 μM Quin-2-AM at 37 °C. Suspensions were then diluted 10-fold in prewarmed medium and incubated for a further 20 min. Cells were then washed three times in medium and incubated at 5 x 10^5/ml in the presence of 75 μM cytochrome *c* prior to the addition of 1 μM fMet–Leu–Phe (A) and 0.1 μg/ml PMA (B). Absorbance changes were then monitored at 550 nm using a Perkin-Elmer Lambda 5 double-beam spectrophotometer, the reference cuvette containing the above constituents plus superoxide dismutase.

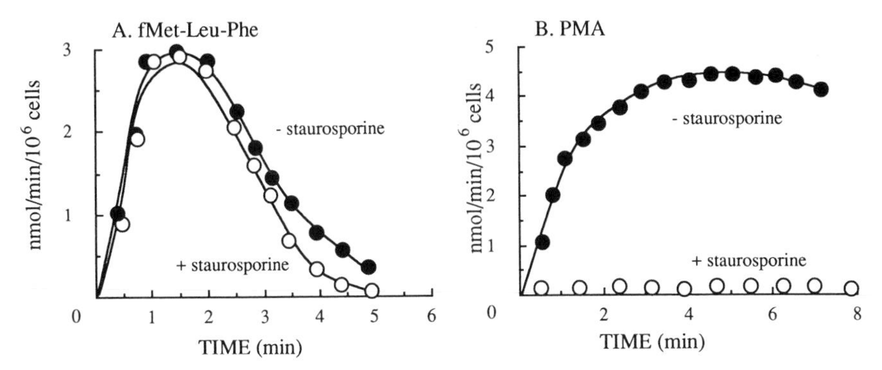

Fig. 4. Effect of staurosporine on O_2^{-} secretion from activated neutrophils. Neutrophils (10^6/ml in RPMI 1640 medium) were incubated for 10 min in the presence (○) or absence (●) of 100 nM staurosporine. Suspensions were then stimulated by the addition of 1 μM fMet–Leu–Phe (A) or 0.1 μg/ml PMA (B) and O_2^{-} secretion measured as described in the legend to Fig. 3. Experiment of F. Watson & J. Robinson.

The fact that phorbol esters activate the respiratory burst and stimulate protein kinase C by substituting for its substrate diacylglycerol (Castagna *et al.*, 1982), is taken as evidence that this kinase is involved in the processes of oxidase activation. It is thus believed to play a role in neutrophil activation by phosphorylation (e.g. at serine and threonine residues) of key oxidase components. Furthermore, inhibition of oxidant production by the protein kinase C inhibitors staurosporine (Tamaoki *et al.*, 1986; Wolf & Baggiolini, 1988; Koenderman *et al.*, 1989; Dewald *et al.*, 1989; Fig. 4) and H-7 (Hidaka *et al.*, 1984; Kawamoto & Hidaka, 1984; Berkow, Dodson & Kraft, 1987) after stimulation by **some** agonists (see below) is taken as further evidence for the role of this enzyme in catalysing the phosphorylation reactions necessary for oxidase activation.

Multiple activation signals

The neutrophil can respond to different specific agonists by occupancy of different receptors which may elicit common cellular responses (e.g. chemotaxis, oxidase activation or degranulation). Hence, it is assumed that receptors which activate identical cellular responses share common signal transduction pathways. However, whilst many of these signal transduction events can be explained by processes outlined in Fig. 1, it is becoming clear that multiple pathways of activation exist in these cells. For example, in cells pre-treated with low concentrations of DAG or PMA, subsequent agonist-mediated activation is uncoupled from intracellular Ca^{2+} transients (Wymann *et al.*, 1987; Baggiolini & Wymann, 1990), and in cells in which Ca^{2+} levels have been increased by ionophore, fMet–Leu–Phe activation occurs in the absence of a further Ca^{2+} rise (Grinstein & Furuya, 1988). Also, 17-hydroxywortmannin (which inhibits receptor-mediated, but not DAG or PMA dependent activation) blocks this Ca^{2+} independent activation (Dewald, Thelan & Baggiolini, 1988): wortmannin itself may inhibit a process leading to phospholipase D activation (Reinhold *et al.*, 1990) which may activate the respiratory burst via its ability to release phosphatidic acid (Bellavite *et al.*, 1988; Bonser *et al.*, 1989). Furthermore, oxidant secretion after exposure to fMet–Leu–Phe (Fig. 4) or opsonised zymosan is unaffected by the protein kinase C inhibitors staurosporine and H-7 (Berkow *et al.*, 1987; Koenderman *et al.*, 1989): PMA-stimulated oxidase activity, unlike that activated by fMet–Leu–Phe is not inhibited by pertussis toxin (Fig. 3), and may be enhanced in Ca^{2+}-buffered cells (Fig. 2).

Hence, whilst intracellular Ca^{2+} transients and protein kinase C activation are undoubtedly involved in oxidase activation, alternative intracellular signal transduction pathways must also exist but as yet these are incompletely defined.

Arachidonic acid and phospholipase A_2

It has been recognised for some time that stimulated neutrophils liberate arachidonic acid via phospholipase A_2 activity, and that this fatty acid may serve as a substrate for cyclooxygenase and lipoxygenase activities during the biosynthesis of prostaglandins and leukotrienes (e.g. LTB_4), respectively. However, several independent lines of evidence also suggest that arachidonic acid and phospholipase A_2 are also involved in the pathways leading to activation of the respiratory burst. Firstly, arachidonic acid (or SDS) is required for oxidase activation (together with the cytosolic components) in the cell free activation system (see later). Secondly, inhibitors of phospholipase A_2 inhibit activated O_2^- generation (Maridonneau-Parini, Tringale & Tauber, 1986; Henderson, Chappell & Jones, 1989) and exogenously-added arachidonic acid can restore activity (Henderson *et al.*, 1989). Thirdly, exogenously-added phospholipase A_2 or arachidonic acid can enhance fMet–Leu–Phe stimulated oxidant production but inhibit PMA stimulated oxidase activity (Lackie & Lawrence, 1987; Lackie, 1988). Thus, it must be concluded that whilst phospholipase A_2 dependent arachidonic acid release is necessary for LTB_4 generation by activated neutrophils, it must also be necessary for oxidase activation acting in conjunction with phospholipases C- and D-generated intracellular signals.

Assembly and activation of the oxidase

A major advance in unravelling the complexities of the molecular pathways necessary for oxidase assembly and activation has been the development of cell-free systems in which $O_2^{\overline{\cdot}}$ generation can be stimulated in broken neutrophil suspensions. Early experiments found that $O_2^{\overline{\cdot}}$ generation could be stimulated in homogenates from unstimulated neutrophils by the addition of anionic detergents such as arachidonic acid or *cis* unsaturated fatty acids (Bromberg & Pick, 1984; Heyneman & Vercauteren, 1984; Curnutte, 1985; McPhail *et al.*, 1985) or SDS (Bromberg & Pick, 1985). Fractionation of these homogenates showed that the plasma membranes and cytosol (but not granules) were required for activation, but whilst Mg^{2+} was essential for activation, neither Ca^{2+} nor ATP (thought to be required for protein kinase C dependent phosphorylations) were necessary (Clark *et al.*, 1987; Curnutte, Kuver & Scott, 1987). Many groups have sought to identify these cytosolic components and here again experiments with neutrophils from patients with CGD have helped in their identification and also to elucidate the molecular processes involved in their assembly. Extracts from patients with CGD could not be activated *in vitro* in these systems, but cytosolic/membrane fractions from X-linked and autosomal recessive patients could be mixed in combination with each other, or else with fractions from normal controls, to restore activity (Curnutte *et al.*, 1987; Curnutte, 1988; Caldwell *et al.*, 1988). For example, as predicted from previous studies, membranes from X-linked CGD patients were defective as they lacked the cytochrome *b*, whereas the cytosolic fraction was defective in most autosomal recessive patients.

Analysis of neutrophil cytosol after Mono Q anion exchange chromatography obtained three active fractions termed NCF-1, -2 and -3 (Nunoi *et al.*, 1988). Independently Volpp, Nauseef & Clark (1988) raised a polyclonal antiserum against the cytosolic components which eluted from a GTP affinity column and this antiserum recognised cytosolic polypeptides of 47 and 65/67 kDa in immunoblots: NCF-1 was shown to contain the 47 kDa polypeptide and NCF-2, the 65/67 kDa component. At present the active components of NCF-3 are yet to be identified. Analysis of defects in the cytosolic components of autosomal recessive patients revealed that most (88%) lacked the 47 kDa component whilst the remainder lacked the 65/67 kDa factor (Clark *et al.*, 1989): both components become associated with the plasma membrane during activation (Clark *et al.*, 1990). cDNA for the 47 kDa has now been cloned and sequenced, and recombinant protein restores oxidase activity in cell free extracts from CGD patients which are defective in this factor (Volpp *et al.*, 1989; Lomax *et al.*, 1989): the derived amino acid sequence data predict a potential phosphorylation site and amino acids 156–221 show some homology to v-*src*, phosphatidylinositol-specific phospholipase C and α-fodrin: a 33 amino acid stretch from 233–265 shares 50% homology to GTPase activating protein. cDNA for the 67 kDa cytosolic factor has also been cloned (Leto *et al.*, 1990): this cDNA encoded a 526 amino acid protein with sequences similar to a motif found in the non-catalytic *src*-related tyrosine kinases.

This 47 kDa polypeptide (or group of related peptides; Segal *et al.*, 1985; Hayakawa *et al.*, 1986; Heyworth & Segal, 1986; Okamura *et al.*, 1988) has been implicated for some time in oxidase function because a band of this relative molecular mass fails to become phosphorylated in most patients with autosomal recessive CGD. This protein is normally phosphorylated in the cytosol and then becomes translocated to the membrane during activation (Kramer *et al.*, 1988; Heyworth, Shrimpton & Segal, 1989), but interestingly, this component cannot be incorporated into membranes of cytochrome b_{-245} deficient CGD neutrophils implying that the 47 kDa phosphoprotein becomes associated with this cytochrome within the membrane (Heyworth *et al.*, 1989).

Oxidase activation – some unsolved problems

The experimental data described above (Fig. 1) provide molecular mechanisms which go some way towards explaining how extracellular signals are coupled to oxidase activation. However, some problems remain in piecing together some of the apparently contradictory observations. For example, it has been shown that during oxidase activation, protein kinase C and components of the oxidase become translocated from the cytosol and specific (perhaps also tertiary) granules, and are incorporated into the plasma membrane. However, experiments with cytoplasts (neutrophils devoid of granules) and cell-free activation systems show that translocation of oxidase components from granules to the plasma membrane is not necessary for activation. Intracellular Ca^{2+} levels increase during agonist-mediated activation, but activation by these agents is not always sensitive to inhibition by protein kinase C inhibitors such as staurosporine (Fig. 4). Phospholipase A_2 (and D) activation occurs during oxidase activation in intact cells, and arachidonic acid stimulates the oxidase in cell free experiments, but their precise roles are obscure. Furthermore, cell-free experiments of oxidase activation require neither added Ca^{2+} nor protein kinase C dependent phosphorylations.

Perhaps some insight into resolving these apparent contradictions may come from two considerations. Firstly, the assembled oxidase on the plasma membrane is rapidly inactivated and new complexes need to be assembled/activated in order to sustain oxidant production (Akard, English & Gabig, 1988). Secondly, one must consider the kinetics of oxidant production elicited by different agents. For example, receptor-mediated activation by small molecules (e.g. fMet–Leu–Phe) begins within seconds after agonist addition and is usually complete within a few minutes: this rapid activation may be protein kinase C/phosphorylation/granule translocation-independent and utilise oxidase components (cytochrome *b* and flavoprotein) already located on the plasma membrane. PMA-activated oxidant production (which begins after a lag phase of a few minutes) requires about 10 min to reach its maximum rate and certainly is protein kinase C/phosphorylation dependent. Perhaps this latter mechanism is necessary to activate and assemble new oxidase complexes (requiring translocation of oxidase components from specific granules) in order to sustain oxidant production as complexes become inactivated. Some evidence in favour of this idea comes from experiments measuring the time course of DAG accumulation after addition of opsonised zymosan and measuring the inhibitory effects of staurosporine on oxygen uptake (Koenderman *et al.*, 1989): DAG accumulation lagged behind oxidant production and only the latter stages of oxygen uptake were inhibited by staurosporine.

Regulation of oxidant production during inflammation

For many years it was believed that mature, circulating neutrophils were terminally-differentiated end of line cells. Being relatively short-lived, it was envisaged that the bloodstream cell had only two fates: either it would respond to a pathogenic challenge by activating its pre-formed molecular apparatus necessary for chemotaxis, phagocytosis and pathogen killing, or else it was cleared from the circulation. *In vitro* evidence loosely supported this notion, in that cells isolated from the circulation could be rapidly activated *in vitro* without any adaptive processes requiring *de novo* biosynthesis, and functions such as phagocytosis and killing were unaffected by inhibitors of transcription and translation (Cline, 1966). Morphological evidence such as poorly-defined endoplasmic reticulum and Golgi apparatus, and few ribosomes, loosely supported this concept of biosynthetic inactivity (Klebanoff & Clark, 1978). Hence, macromolecular biosynthesis was not considered necessary nor even possible in these cells, and the idea that they could respond to inflammatory signals in their environment by a molecular re-organisation was thought equally unlikely.

It was therefore of considerable interest and of great importance when it was discovered that neutrophils isolated from the bloodstream of patients with acute bacterial infections had greater oxidative activity than those from a control group (McCall *et al.*, 1979). The possibility that this was due to release into the bloodstream of a functionally more active sub-population of neutrophils was excluded when it was found that bloodstream cells from healthy controls could generate greater levels of oxidants upon subsequent stimulation if they were first pre-incubated with sub-stimulatory levels of agonists or chemoattractants (Van Epps & Garcia, 1980; Bender, McPhail & Van Epps, 1983; McPhail, Clayton & Snyderman, 1984; Dewald & Baggiolini, 1985). Hence, the idea that circulating cells could be primed *in vivo* and *in vitro* was proposed: primed neutrophils thus generate greater levels of oxidants and are more bactericidal, fungicidal and tumouricidal than unprimed cells. It was later shown that cytokines such as γ-interferon (Berton *et al.*, 1986) could also prime (but not activate) oxidant production and bactericidal activity (Edwards, Say & Hughes, 1988) and because such cytokines are generated during infections or inflammatory diseases by activated immune cells, this is likely to be a mechanism of regulating neutrophil function *in vivo*.

Molecular mechanisms underlying priming

Priming of oxidant production after stimulation by fMet–Leu–Phe is observed after pre-incubation of bloodstream neutrophils with γ-interferon (Edwards *et al.*, 1988) and GM-CSF (Edwards *et al.*, 1989) for periods in excess of 15 min. Interestingly, oxidant production stimulated by PMA (which by-passes receptor occupancy/signal transduction pathways) is not increased during priming, although the lag time before oxidase activation occurs may be decreased, suggesting that a partial assembly/activation of the oxidase occurs during priming. The rapid up-regulation of receptor-mediated activation is explained, at least in part, by increased expression of some plasma membrane receptors, for example those for fMet–Leu–Phe (Weisbart, Golde & Gasson, 1986) and Mac-1 (Arnaout *et al.*, 1986), or else by changes in the affinities of existing receptors (Weisbart *et al.*, 1986; Weisbart, 1989). Because such receptor up-regulation is very rapid and independent of protein biosynthesis (Edwards *et al.*, 1990; Fig. 5), it is likely to arise from the mobilisation of pre-existing sub-cellular pools of receptors which become translocated to the plasma membrane.

However, a role for *de novo* protein biosynthesis in cytokine-dependent priming was shown when it was discovered that the enhanced oxidant production induced during a 4 h incubation with γ-interferon was

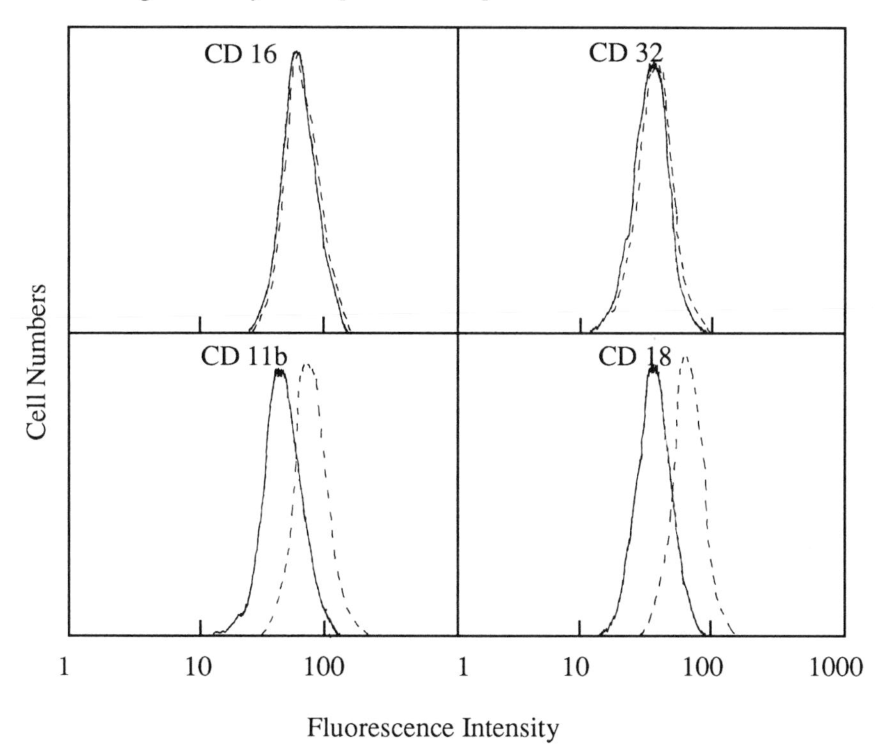

Fig. 5. Effect of GM-CSF on receptor expression. Neutrophils (10^7/ml) were incubated for 15 min in the presence (---) and absence(——) of 50 u/ml rGM-CSF. Expression of CD16(FcRIII), CD32(FcRII) and CD 11b/CD18(Mac/1) was then determined by FACS analysis as described in Edwards *et al.* (1990).

blocked by cycloheximide (Berton *et al.*, 1986). This observation raised the possibility that cytokine treatment up-regulated macromolecular biosynthesis in these cells and re-opened the question as to the role of biosynthesis in neutrophil function during inflammation.

Gene expression in bloodstream neutrophils

The concept that the mature neutrophil was biosynthetically inert contrasted with our understanding of monocyte function, where it has long been recognised that active gene expression plays an important role in the function of these cells during inflammatory activation. This view of the neutrophil was sustained in spite of the fact that early experiments directly measuring the rates of incorporation of radiolabelled RNA- and protein-precursors into circulating neutrophils showed that macromolecular biosynthesis does indeed occur (albeit at fairly low rates) in these cells (Granelli-Piperno, Vassalli & Reich, 1977; 1979). For example, it was shown that bloodstream neutrophils actively-synthesise plasminogen activator and that biosynthesis of this and other polypeptides is regulated by corticosteroids (Granelli-Piperno *et al.*, 1977; Blowers, Jayson & Jasani, 1985; 1988). The picture which is now gradually emerging is that of a neutrophil sharing many of the biosynthetic and immuno-regulatory functions previously thought to be restricted to monocytes: thus, neutrophils actively synthesise essential components necessary to sustain their own function in response to signals in their environment, and they also synthesise immuno-regulatory components which can direct the progress of an inflammatory response. In view of the potential pharmacological and pathological importance of these observations, it is somewhat surprising that this phenomenon has until recently been overlooked.

We have shown that bloodstream neutrophils actively synthesise a variety of polypeptides (Hughes, Humphreys & Edwards, 1987; Humphreys, Hughes & Edwards, 1989; Edwards *et al.*, 1989) which can be visualised by 2D-PAGE (Fig. 6). Experiments using antibodies to immunoprecipitate radiolabelled constituents, or gene probes to detect levels of particular mRNAs have shown that bloodstream neutrophils actively synthesise: actin and some plasma membrane components (Jack & Fearon, 1988; Jost *et al.*, 1990); cytochrome b_{-245} of the NADPH oxidase (Cassatella *et al.*, 1989); several stress proteins (Maridonneau-Parini, Clerc & Polla, 1988) and c-*fos* (Colotta *et al.*, 1987). Because many neutrophil functions such as chemotaxis, adhesion and opsonophagocytosis are receptor-mediated, it is of interest to note that a number of receptors or elements of the cytoskeletal system are actively synthesised by bloodstream cells. This biosynthesis may be necessary to replenish and cycle receptors which may be lost due to shedding or membrane turnover, and hence ensure adequate levels of expression necessary to maintain functional responsiveness. We decided, therefore, to determine the relationship between receptor expression and oxidant production, and to establish the importance of macromolecular biosynthesis on the maintenance of receptor-mediated functions.

Neutrophils were incubated at 37 °C for periods of up to 5 h and then tested for their ability to generate reactive oxidants in response to fMet–Leu–Phe and PMA: in order to determine the importance of protein biosynthesis in these processes, suspensions were also incubated with cycloheximide. In control suspensions the ability to generate reactive oxidants only declined by about 20% over a 5 h incubation period, but in cycloheximide-treated suspensions the ability to generate oxidants began

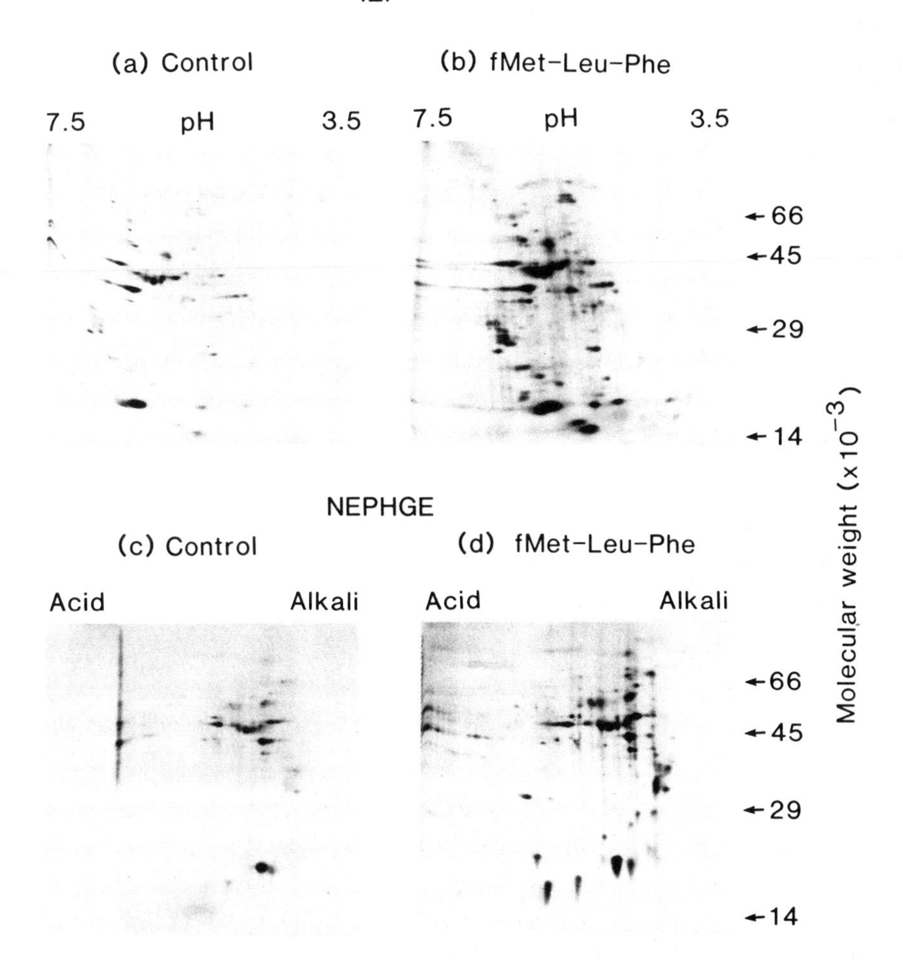

Fig. 6. Two-dimensional polyacrylamide gel electrophoresis (2D-PAGE) of newly-synthesised neutrophil proteins. Neutrophils were incubated in RPMI 1640 medium supplemented with 60 μCi/ml [^{35}S]-methionine in the absence (*a*, *c*) or presence (*b*, *d*) of 0.1 μM fMet–Leu–Phe. After 60 min incubation at 37 °C radiolabelled polypeptides were precipitated with 10% TCA (final conc) and analysed by 2D-PAGE using either IEF (isoelectric focussing) or NEPHGE (non-equilibrium pH gel electrophoresis) in the first dimension, prior to separation in a 13% polacrylamide gel (plus SDS) in the second dimension. Reproduced from Hughes *et al.* (1987).

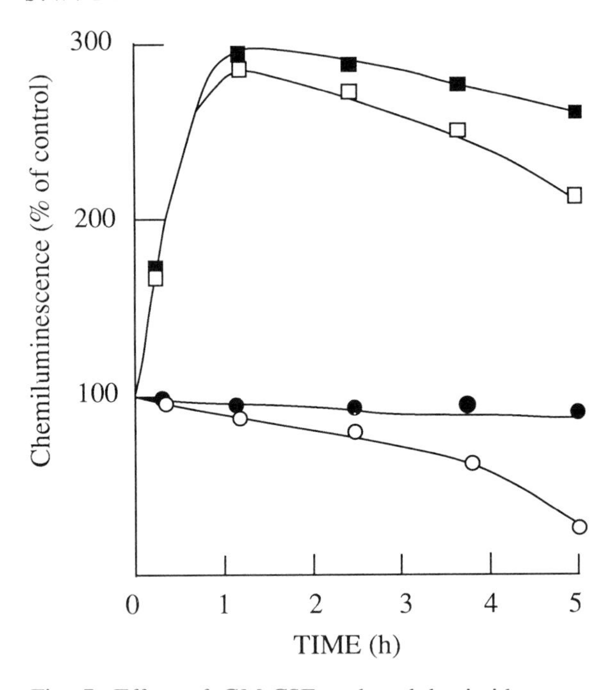

Fig. 7. Effect of GM-CSF and cycloheximide on neutrophil oxidant production. Neutrophils (10^7/ml in RPMI 1640 medium) were incubated at 37 °C in the absence (●,○) and presence (■, □) of 50 u/ml rGM-CSF and the presence (○,□) and absence (●,■) of 30 μg/ml cycloheximide. At time intervals, aliquots of 10^6 cells were removed, luminol added to 10 μM (final conc, total volume 1 ml) and the chemiluminescence after the addition of 1 μM fMet–Leu–Phe recorded.

to decline by 2 h incubation and by 5 h was only about 25% of that of the untreated suspension (Fig. 7). Thus, it appears that *de novo* protein biosynthesis is necessary to **maintain** the ability of these cells to generate oxidants in response to receptor-mediated activation. However, when PMA-stimulated oxidase activity was measured in these suspensions, the response was not as sensitive to inhibition by cycloheximide as the fMet–Leu–Phe response: PMA-stimulated oxidase activity required longer incubations with cycloheximide before a reduction was observed and at all time points examined the responses to PMA were always more preserved than the fMet–Leu–Phe response, e.g. at 4 h incubation in the presence of cycloheximide PMA-induced oxidase activity was reduced by 10% (±11%, n=6), whereas that activated by fMet–Leu–Phe was inhibited by 50% (±10%, n=8). Thus, these observations suggest that *de*

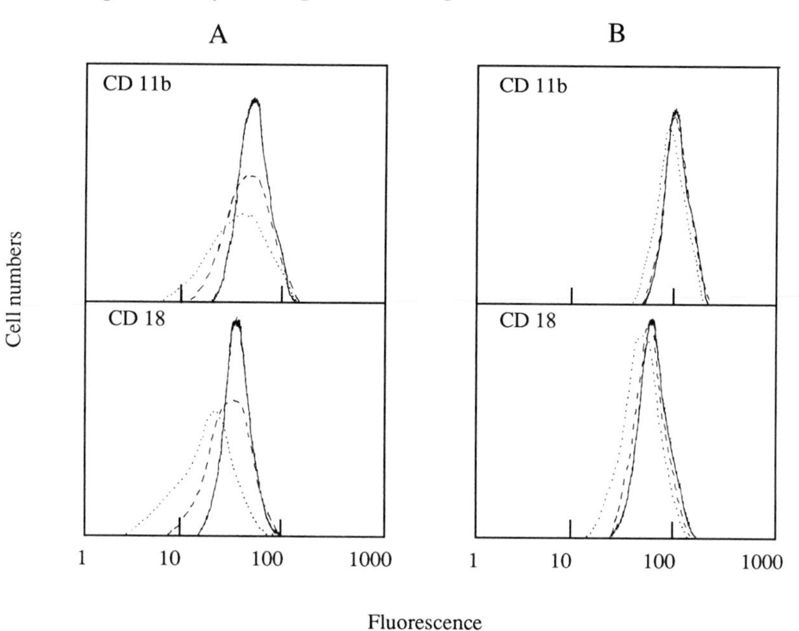

Fig. 8. Effect of GM-CSF on receptor expression. Neutrophils were incubated as described in the legend to Fig. 7 in the absence (A) and presence (B) of 50 u/ml rGM-CSF. At time intervals of 15 min (——), 3 h (---) and 5 h (·····), aliquots were removed and the expression of CD11b and CD18 measured by FACS analysis.

novo protein biosynthesis is necessary to maintain oxidase function and that receptor-mediated activation pathways are more sensitive to the inhibitory effects of cycloheximide, indicating a greater dependency on *de novo* biosynthesis.

Therefore, in order to determine if macromolecular biosynthesis was important for the maintenance of receptor expression, FACS analysis was used to measure the expression of CD11b/CD18 (Mac-1, the C3bi receptor), CD32 (FcRII) and CD16 (FcRIII) as cells aged or were blocked in biosynthesis. Expression of CD32 and CD16 was well maintained during incubation of control suspensions for up to 5 h, but expression of CD11b/18 'broadened' somewhat as the cells aged (Fig. 8). This indicated an increased heterogeneity in the levels of expression of these epitopes as cells aged in culture. In suspensions blocked in protein biosynthesis by cycloheximide, expression of some receptors was markedly reduced. For example, expression of CD18 and CD16 was greatly

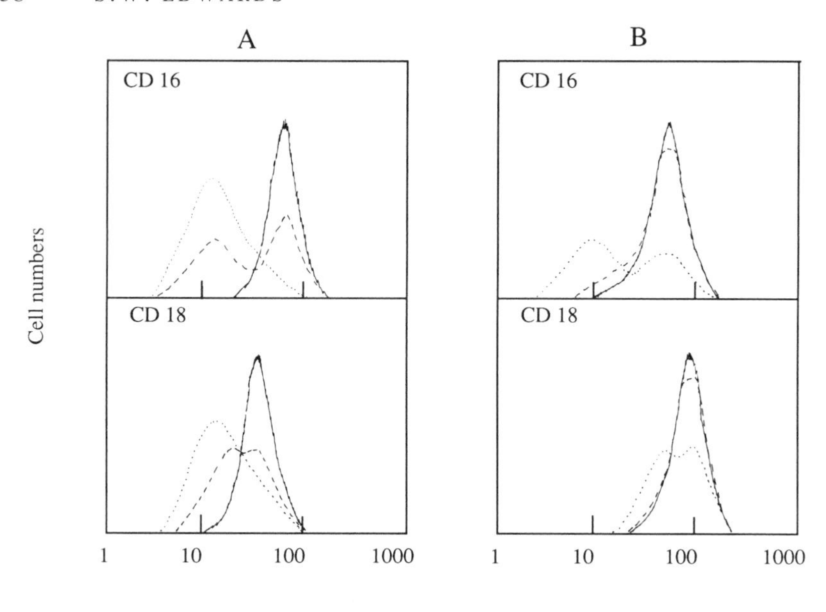

Fig. 9. Effect of cycloheximide on receptor expression. Neutrophils were incubated as described in the legend to Fig. 7, in the presence of 30 μg/ml cycloheximide. At 15 min (——), 3 h (---) and 5 h (·····) incubation in the absence (A) and presence (B) of 50 u/ml rGM-CSF aliquots were removed and the expression of CD16 and CD18 measured by FACS analysis.

impaired (Fig. 9) and two sub-populations of normal and low receptor expression cells were visualised in cycloheximide-treated suspensions.

Hence, it may be concluded from these experiments that active protein biosynthesis is necessary for the maintenance of neutrophil responsiveness, and that continued expression of some membrane receptors and ability to activate the NADPH oxidase is also dependent upon this phenomenon. This is likely to be of pathological importance as impairment of biosynthesis (either as a result of disease or its therapy) will severely impair neutrophil function and hence interfere with their ability to mount an acute inflammatory response.

Effects of cytokines on gene expression

The fact that cycloheximide prevented the up-regulation of neutrophil function observed during a 4 h incubation with γ-interferon (Berton *et al.*,

1986), suggested that cytokines may activate gene expression in these cells. Furthermore, whilst fibronectin is not synthesised in bloodstream neutrophils (Marino, Davis & Spagnuolo, 1987) it is actively synthesised in cells isolated from the synovial fluid of patients with rheumatoid arthritis (Beaulieu *et al.*, 1987) which is known to contain several cytokines (Yamagata *et al.*, 1988). Hence, the possibility that cytokine exposure selectively activates gene expression in neutrophils was tested by analysis of polypeptides labelled by [^{35}S]-methionine after incubation in the presence and absence of cytokines or other priming agents. Hence, we have shown that exposure of bloodstream neutrophils to fMet–Leu–Phe (Hughes *et al.*, 1987), γ-interferon (Humphreys *et al.*, 1989), GM-CSF (Edwards *et al.*, 1989) and G-CSF (in preparation) results in a 2–4-fold increase in the rate of protein biosynthesis. Furthermore, analysis of newly-labelled polypeptides by 2D-PAGE showed that two classes of polypeptides were identified (Fig. 6): the rate of labelling of one of these changed very little (1–3-fold), whereas the rate of labelling of the other class increased 5–20-fold upon cytokine exposure. We have proposed that these cytokine-regulated genes play an important role in neutrophil function in inflammation because neutrophil function is regulated by cytokines *in vivo*. It has recently been shown that cytokine-exposed neutrophils themselves secrete secondary cytokines such as IL-1 (Lindemann *et al.*, 1988), G-CSF, M-CSF (Lindemann *et al.*, 1989) and α-interferon (Shirafuji *et al.*, 1990) and there is some evidence to suggest that inflammatory neutrophils also express TNF (Yamazaki, Ikenami & Sugiyama, 1989).

Therefore, the effects of cytokines such as GM-CSF on the up-regulation and maintenance of receptor expression were investigated and correlated with the ability of this agent to enhance respiratory burst activity. After incubation of neutrophils with GM-CSF for 15 min, there was considerable up-regulation of Mac-1 (CD11b/CD18) expression as assessed by FACS analysis, but little change in the expression of CD32 and CD16 (Fig. 5). This was accompanied by an increased ability to mount a respiratory burst which was maximally up-regulated after 1 h incubation with this cytokine (Fig. 7). Both of these phenomena were **independent** of *de novo* protein biosynthesis (i.e. unaffected by cycloheximide) confirming the idea that receptor up-regulation results from mobilisation of pre-existing sub-cellular receptor pools. However, GM-CSF also prevented the decay in receptor expression as neutrophils aged in culture: fluorescence distributions for CD11b and CD18 were better preserved in GMCSF-exposed suspensions than in controls (Fig. 8). This may be related to the ability of GM-CSF to extend the functional lifetime of neutrophils (Lopez *et al.*, 1986). However, this enhancement of receptor expression was also dependent upon *de novo* protein biosynthesis as this also was blocked during incubation with cycloheximide (Fig. 9).

Conclusions and perspective

Neutrophils possess a $O_2^{\cdot-}/H_2O_2$-generating NADPH oxidase which is utilised for pathogen killing and its inappropriate activation may result in the secretion of tissue-damaging oxidants in inflammatory conditions. Hence, complex mechanisms exist to activate and deactivate this oxidase upon receiving appropriate signals from the extracellular environment of the neutrophil. Neutrophil function *in vivo* is therefore regulated very closely by the expression of specific receptors which are coupled to intracellular signalling pathways to activate the appropriate cellular response.

The discoveries that cytokines can up-regulate neutrophil function and that these cells have the capacity for active biosynthesis has completely changed our understanding of how these cells function during inflammatory activation. Cytokine-exposure leads to a rapid increase in the expression of some plasma membrane receptors and primes the respiratory burst to generate elevated levels of oxidants: cytokines also activate a selective increase in neutrophil gene expression which is necessary to augment neutrophil function, and also extend the functional lifespan of these cells. We are only beginning to identify these cytokine-regulated gene products, but these include secondary cytokines and other important components which are necessary either to sustain neutrophil function or else augment the inflammatory response by acting upon other immune cells.

By virtue of their ability to secrete secondary cytokines, the neutrophil must now be considered to be an active immuno-regulatory cell, sharing abilities which enable it to direct the progress of an inflammatory response in a similar way to macrophages and lymphocytes. Further work over the next decade will undoubtedly discover the full biosynthetic capacity of these cells and will identify their full role in inflammatory activation.

References

Akard, L.P., English, D. & Gabig, T.G. (1988). Rapid deactivation of NADPH oxidase in neutrophils: continuous replacement by newly activated enzyme sustains the respiratory burst. *Blood* **72**, 322–7.

Al-Mohanna, F.A. & Hallett, M.B. (1988). The use of fura-2 to determine the relationship between cytoplasmic free Ca^{2+} and oxidase activation in rat neutrophils. *Cell Calcium* **9**, 17–26.

Anderson, C.L. & Abraham, G.N. (1980). Characterisation of the Fc receptor for IgG on a human macrophage cell line, U937, *Journal of Immunology* **125**, 2735–41.

Anderson, C.L. & Looney, R.J. (1986). Human leukocyte Ig Fc receptors. *Immunology Today* **7**, 264–6.

Anderson, D.C. & Springer, T.A. (1987). Leukocyte adhesion deficiency: an inherited defect in the Mac-1, LFA-1 and p150,95 glycoproteins. *Annual Reviews of Medicine* **38**, 175–94.

Anderson, D.C., Schmalsteig, F.C., Finegold, M.J., Hughes, B.J., Rothlein, R., Miller, L.J., Kohl, S., Tosi, M.F., Jacobs, R., Waldrop, T.C., Goldman, A.S., Shearer, W.T. & Springer, T.A. (1985). The severe and moderate phenotypes of heritable Mac-1, LFA-1, p150,95 deficiency: their quantitative definition and relation to leukocyte dysfunction and clinical features. *Journal of Infectious Diseases* **152**, 668–89.

Arnaout, M.A. (1990). Structure and function of the leukocyte adhesion molecules CD11/CD18. *Blood* **75**, 1037–50.

Arnaout, M.A., Spits, H., Terhorst, C., Pitt, J. & Todd, R.F.III. (1984). Deficiency of a leukocyte surface glycoprotein (LFA-1) in two patients with Mo1 deficiency. Effects of cell activation on Mol/LFA-1 surface expression in normal and deficient leukocytes. *Journal of Clinical Investigation* **74**, 1291–300.

Arnaout, M.A., Wang, E.A., Clark, S.C. & Sieff, C.A. (1986). Human recombinant granulocyte-macrophage colony-stimulating factor increases cell-to-cell adhesion and surface expression of adhesion-promoting surface glycoproteins on mature granulocytes. *Journal of Clinical Investigation* **78**, 597–601.

Babior, B.M. (1978). Oxygen-dependent microbial killing by phagocytes. *New England Journal of Medicine* **298**, 659–68.

Babior, B.M. (1984*a*). Oxidants from phagocytes: agents of defense and destruction. *Blood* **64**, 959–66.

Babior, B.M. (1984*b*). The respiratory burst of phagocytes. *Journal of Clinical Investigation* **73**, 599–601.

Babior, B.M. (1988). The respiratory burst oxidase. In *Phagocytic defects II. Hematology/Oncology Clinics of North America*, ed. J.T. Curnutte, pp. 201–12. Philadelphia: Saunders.

Babior, B.M., Kipnes, R.S. & Curnutte, J.T. (1973). Biological defense mechanisms. The production by leukocytes of superoxide, a potential bacteriocidal agent. *Journal of Clinical Investigation* **52**, 741–4.

Baggiolini, M. & Wymann, M.P. (1990). Turning on the respiratory burst. *Trends in Biochemical Sciences* **15**, 69–72.

Baldridge, C.W. & Gerard, R.W. (1933). The extra respiration of phagocytosis. *American Journal of Physiology* **103**, 235–6.

Bannister, J.V. & Bannister, W.H. (1985). Production of oxygen-centered radicals by neutrophils and macrophages as studied by electron spin resonances (ESR). *Environmental Health Perspectives* **64**, 37–43.

Beaulieu, A.D., Lang, F., Belles-Isles, M. & Poubelle, P. (1987). Protein biosynthetic activity of polymorphonuclear leukocytes in inflammatory arthropathies. Increased synthesis and release of fibronectin. *Journal of Rheumatology* **14**, 656–61.

Bellavite, P., Corso, F., Dusi, S., Grzeskowiak, M., Della-Bianca, V. & Rossi, F. (1988). Activation of NADPH-dependent superoxide production in plasma membrane extracts of pig neutrophils by phosphatidic acid. *Journal of Biological Chemistry* **263**, 8210–14.

Beller, D.I., Springer, T.A. & Schreiber, R.D. (1982). Anti-Mac-1 selectively inhibits the mouse and human type three complement receptor. *Journal of Experimental Medicine* **156**, 1000–9.

Bender, J.G., McPhail, L.C. & Van Epps, D.E. (1983). Exposure of human neutrophils to chemotactic factors potentiates activation of the respiratory burst enzyme. *Journal of Immunology* **130**, 2316–23.

Berger, M., O'Shea, J., Cross, A.S., Folks, T.M., Chused, T.M., Brown, E.J. & Frank, M.M. (1984). Human neutrophils increase expression of C3bi as well as C3b receptors upon activation. *Journal of Clinical Investigation* **74**, 1566–71.

Berkow, R.L., Dodson, R.W. & Kraft, A.B. (1987). The effect of a protein kinase C inhibitor, H-7, on human neutrophil oxidative burst and degranulation. *Journal of Leukocyte Biology* **41**, 441–6.

Berridge, M.J. & Irvine, R.F. (1984). Inositol triphosphate, a novel second messenger in cellular signal transduction. *Nature* **312**, 315–21.

Berridge, M.J. & Irvine, R.F. (1989). Inositol phosphates and cell signalling. *Nature* **341**, 197–204.

Berton, G., Zeni, L., Cassatella, M.A. & Rossi, F. (1986). Gamma interferon is able to enhance the oxidative metabolism of human neutrophils. *Biochemical and Biophysial Research Communications* **138**, 1276–82.

Blowers, L.E., Jayson, M.I.V. & Jasani, M.K. (1985). Effect of dexamethasone on polypeptides synthesized in polymorphonuclear leucocytes. *FEBS Letters* **181**, 362–6.

Blowers, L.E., Jayson, M.I.V. & Jasani, M.K. (1988). Dexamethasone modulated protein synthesis in polymorphonuclear leukocytes: response in rheumatoid arthritis. *Journal of Rheumatology* **15**, 785–90.

Bohler, M.-C., Seger, R.A., Mouy, R., Vilmer, E., Fischer, A. & Griscelli, C. (1986). Study of 25 patients with chronic granulomatous disease: a new classification by correlating respiratory burst, cytochrome b and flavoprotein. *Journal of Clinical Immunology* **6**, 136–45.

Bonser, R.W., Thompson, N.T., Randall, R.W. & Garland, L.G. (1989). Phospholipase D activation is functionally linked to superoxide generation in the human neutrophil. *Biochemical Journal* **264**, 617–20.

Borregaard, N., Cross, A.R., Herlin, T., Jones, O.T.G., Segal, A.W. & Valerius, N.H. (1983*b*). A variant form of X-linked chronic granulomatous disease with normal nitroblue tetrazolium slide test and cytochrome b. *European Journal of Clinical Investigation* **13**, 243–7.

Borregaard, N., Heiple, J.M., Simons, E.R. & Clark, R.A. (1983*a*). Subcellular localisation of the *b*-cytochrome component of the human microbicidal oxidase: translocation during activation. *Journal of Cell Biology* **97**, 52–61.

Borregaard, N. & Tauber, A.I. (1984). Subcellular localisation of the human neutrophil NADPH oxidase: *b*-cytochrome and associated flavoprotein. *Journal of Biological Chemistry* **259**, 47–52.

Boxer, L.A. & Smolen, J.E. (1988). Neutrophil granule constituents and their release in health and disease. In *Phagocyte Defects I. Hematology/Oncology Clinics of North America*, ed. J.T. Curnutte, pp. 101–34. Philadelphia: Saunders.

Britigan, B.E., Cohen, M.S. & Rosen, G.M. (1987). Detection of the production of oxygen-centered free radicals by human neutrophils using spin trapping techniques: a critical perspective. *Journal of Leukocyte Biology* **41**, 349–62.

Britigan, B.E., Hassett, D.J., Rosen, G.M., Hamill, D.R. & Cohen, M.S. (1989). Neutrophil degranulation inhibits potential hydroxyl-radical formation. Relative impact of myeloperoxidase and lactoferrin release on hydroxyl-radical production by iron-supplemented neutrophils assessed by spin-trapping techniques. *Biochemical Journal* **264**, 447–55.

Bromberg, Y. & Pick, E. (1984). Unsaturated fatty acids stimulate NADPH-dependent superoxide production in a cell-free system derived from macrophages. *Cellular Immunology* **88**, 213–21.

Bromberg, Y. & Pick, E. (1985). Activation of NADPH-dependent superoxide production in a cell free system by sodium dodecyl sulfate. *Journal of Biological Chemistry* **260**, 13539–45.

Caldwell, S.E., McCall, C.E., Hendricks, C.E., Leone, P.A., Bass, D.A. & McPhail, L.C. (1988). Coregulation of NADPH oxidase activation and phosphorylation of a 48-kD protein(s) by a cytosolic factor defective in autosomal recessive chronic granulomatous disease. *Journal of Clinical Investigation* **81**, 1485–96.

Cassatella, M.A., Hartman, L., Perussia, B. & Trinchieri, G. (1989). Tumour necrosis factor and immune interferon synergistically induce cytochrome b_{-245} heavy-chain gene expression and nicotinamide-adenine dinucleotide phosphate hydrogenase oxidase in human leukemic myeloid cells. *Journal of Clinical Investigation* **83**, 1570–9.

Castagna, M., Takai, Y., Kaibuchi, K., Sano, K., Kikkawa, U. & Nishizuka, Y. (1982). Direct activation of calcium-activated, phospholipid-dependent protein kinase C by tumor-promoting phorbol esters. *Journal of Biological Chemistry* **257**, 7847–51.

Clark, R.A., Leidel, K.G., Pearson, D.W. & Nauseef, W.M. (1987).

NADPH oxidase of human neutrophils. Subcellular localization and characterization of an arachidonate-activatable superoxide-generating system. *Journal of Biological Chemistry* **262**, 4065–74.

Clark, R.A., Malech, H.L., Gallin, J.I., Nunoi, H., Volpp, B.D., Pearson, D.W., Nauseef, W.M. & Curnutte, J.T. (1989). Genetic variants of chronic granulomatous disease: prevalence of deficiencies of two cytosolic components of the NADPH oxidase. *New England Journal of Medicine* **312**, 647–52.

Clark, R.A., Volpp, B.D., Leidal, K.G. & Nauseef, W.M. (1990). Two cytosolic components of the human neutrophil respiratory burst oxidase translocate to the plasma membrane during cell activation. *Journal of Clinical Investigation* **85**, 714–21.

Cline, M.J. (1966). Phagocytosis and synthesis of ribonucleic acid in human granulocytes. *Nature* **212**, 1431–3.

Cohen, M.S., Britigan, B.E., Hassett, D.J. & Rosen, G.M. (1988). Phagocytes, O_2 reduction, and hydroxyl radical. *Reviews of Infectious Diseases* **10**, 1088–96.

Colotta, F., Wang, J.M., Polentarutti, N. & Mantovani, A. (1987). Expression of c-*fos* protooncogene in normal human peripheral blood granulocytes. *Journal of Experimental Medicine* **165**, 1224–9.

Crawford, D.R. & Schneider, D.L. (1982). Identification of ubiquinone-50 in human neutrophils and its role in microbicidal events. *Journal of Biological Chemistry* **257**, 6662–8.

Cross, A.R., Higson, F.K., Jones, O.T.G., Harper, A.M. & Segal, A.W. (1982*a*). The enzymic reduction and kinetics of oxidation of cytochrome b_{-245} of neutrophils. *Biochemical Journal* **204**, 479–85.

Cross, A.R. & Jones, O.T.G. (1986). The effect of the inhibitor diphenylene iodonium on the superoxide-generating system of neutrophils. Specific labelling of a component of the oxidase. *Biochemical Journal* **237**, 111–16.

Cross, A.R., Jones, O.T.G., Garcia, R. & Segal, A.W. (1982*b*). The association of FAD with the cytochrome b_{-245} of human neutrophils. *Biochemical Journal* **208**, 759–63.

Cross, A.R., Jones, O.T.G., Harper, A.M. & Segal, A.W. (1981). Oxidation-reduction properties of the cytochrome *b* found in the plasma membrane fraction of human neutrophils. *Biochemical Journal* **194**, 599–606.

Cross, A.R., Parkinson, J.F. & Jones, O.T.G. (1985). Mechanisms of the superoxide producing oxidase of neutrophils. O_2 is necessary for the fast reduction of cytochrome b_{-245} by NADPH. *Biochemical Journal* **226**, 881–4.

Cunningham, C.C., DeChatelet, L.R., Spach, P.I., Parce, J.W., Thomas, M.J., Lees, C.J. & Shirley, P.S. (1982). Identification and quantification of electron-transport components in human polymorphonuclear neutrophils. *Biochemica et Biophysia Acta* **682**, 430–5.

Curnutte, J.T. (1985). Activation of human neutrophil nicotinamide adenine dinucleotide phosphate, reduced (triphosphopyridine nucleotide, reduced) oxidase by arachidonic acid in a cell-free system. *Journal of Clinical Investigation* **75**, 1740–3.

Curnutte, J.T. (ed.) (1988). *Phagocytic Defects I and II. Hematology/Oncology Clinics of North America,* p. 336. Philadelphia: Saunders.

Curnutte, J.T., Kuver, R. & Scott, P.J. (1987). Activation of neutrophil NADPH oxidase in a cell-free system. Partial purification of components and characterisation of the activation process. *Journal of Biological Chemistry* **262**, 5563–9.

Dewald, B. & Baggiolini, M. (1985). Activation of NADPH oxidase in human neutrophils. Synergism between fMLP and the neutrophil products PAF and LTB_4. *Biochemical and Biophysical Research Communications* **128**, 297–304.

Dewald, B., Thelan, M. & Baggiolini, M. (1988). Two transduction sequences are necessary for neutrophil activation by receptor agonists. *Journal of Biological Chemistry* **263**, 16179–84.

Dewald, B., Thelan, M., Wymann, M.P. & Baggiolini, M. (1989). Staurosporine inhibits the respiratory burst and induces exocytosis in human neutrophils. *Biochemical Journal* **264**, 879–84.

Dinauer, M.C., Curnutte, J.T., Rosen, H. & Orkin, S.H. (1989). A missense mutation in the neutrophil cytochrome *b* heavy chain in cytochrome-positive X-linked chronic granulomatous disease. *Journal of Clinical Investigation* **84**, 2012–16.

Dinauer, M.C., Orkin, S.H., Brown, R., Jesaitis, A.J. & Parkos, C.A. (1987). The glycoprotein encoded by the X-linked chronic granulomatous disease locus is a component of the neutrophil cytochrome *b* complex. *Nature* **327**, 717–20.

Doussiere, J. & Vignais, P.V. (1985). Purification and properties of an $O_2^{\overline{\cdot}}$-generating oxidase from bovine polymorphonuclear leukocytes. *Biochemistry* **24**, 7231–9.

Edwards, S.W., Holden, C.S., Humphreys, J.M. & Hart, C.A. (1989). Granulocyte-macrophage colony-stimulating factor (GM-CSF) primes the respiratory burst and stimulates protein biosynthesis in human neutrophils. *FEBS Letters* **256**, 62–6.

Edwards, S.W. & Lloyd, D. (1986). Formation of myeloperoxidase compound II during aerobic stimulation of rat neutrophils. *Bioscience Reports* **6**, 275–82.

Edwards, S.W. & Lloyd, D. (1987). CO-reacting haemoproteins of neutrophils: evidence for cytochrome b-245 and myeloperoxidase as potential oxidases during the respiratory burst. *Bioscience Reports* **7**, 193–9.

Edwards, S.W., Say, J.E. & Hughes, V. (1988). Gamma interferon enhances the killing of *Staphylococcus aureus* by human neutrophils. *Journal of General Microbiology* **134**, 37–42.

Edwards, S.W., Watson, F., MacLeod, R. & Davies, J.M. (1990).

Receptor expression and oxidase activity in human neutrophils: regulation by granulocyte-macrophage colony-stimulating factor and dependence upon protein biosynthesis. *Bioscience Reports* **10**, 393–401.

Ellis, J.A., Cross, A.R. & Jones, O.T.G. (1989). Studies on the electron-transfer mechanism of the human neutrophil NADPH oxidase. *Biochemical Journal* **262**, 575–9.

Elsbach, P. & Weiss, J. (1983). A reevaluation of the role of the O_2-dependent and O_2-independent microbial systems of phagocytes. *Reviews of Infectious Diseases* **5**, 843–53.

Gabig, T.G. (1983). The NADPH dependent O_2^{-} generating oxidase from human neutrophils. Identification of a flavoprotein component that is deficient in a patient with chronic granulomatous disease. *Journal of Biological Chemistry* **258**, 6352–6.

Gabig, T.G., English, D., Akard, L.P. & Schell, M.J. (1987). Regulation of neutrophil NADPH oxidase activation in a cell-free system by guanine nucleotides and fluoride: evidence for participation of a pertussis and cholera toxin-sensitive G protein. *Journal of Biological Chemistry* **262**, 1685–90.

Gabig, T.G. & Lefker, B.A. (1984). Deficient flavoprotein component of the NADPH-dependent O_2^{-}-generating oxidase in the neutrophils from three male patients with chronic granulomatous disease. *Journal of Clinical Investigation* **73**, 701–5.

Gabig, T.G. & Lefker, B.A. (1985). Activation of the human neutrophil NADPH oxidase results in coupling of electron carrier function between ubiquinone-10 and cytochrome b_{559}. *Journal of Biological Chemistry* **260**, 3991–5.

Ganz, T., Selsted, M.E., Szklarek, D., Harwig, S.S.L., Daher, K., Bainton, D.F. & Lehrer, R.I. (1985). Defensins: natural peptide antibiotics of human neutrophils. *Journal of Clinical Investigation* **76**, 1427–35.

Garcia, R.C. & Segal, A.W. (1984). Changes in the subcellular distribution of the cytochrome b_{-245} on stimulation of human neutrophils. *Biochemical Journal* **219**, 233–42.

Garcia, R.C. & Segal, A.W. (1988). Phosphorylation of the subunits of cytochrome b_{-245} upon triggering of the respiratory burst of human neutrophils and macrophages. *Biochemical Journal* **252**, 901–4.

Granelli-Piperno, A., Vassalli, J.D. & Reich, E. (1977). Secretion of plasminogen activator by human polymorphonuclear leukocytes. Modulation by glucocorticoids and other effectors. *Journal of Experimental Medicine* **146**, 1693–1706.

Granelli-Piperno, A., Vassalli, J.D. & Reich, E. (1979). RNA and protein synthesis in human peripheral blood polymorphonuclear leukocytes. *Journal of Experimental Medicine* **149**, 284–9.

Grinstein, S. & Furuya, W. (1988). Receptor-mediated activation of electropermealised neutrophils. Evidence for a Ca^{2+}- and protein kinase C-independent signalling pathway. *Journal of Biological Chemistry* **263**, 1779–83.

Hallett, M.B. & Campbell, A.K. (1983). Two distinct mechanisms for stimulation of oxygen radical production by polymorphonuclear leukocytes. *Biochemical Journal* **216**, 459–65.

Hallett, M.B., Edwards, S.W. & Campbell, A.K. (1987). Control of oxygen radical production by polymorphonuclear leukocytes monitored by luminol-dependent chemiluminescence: the roles of intracellular Ca^{2+}, oxygen and redox components. In *Cellular Chemiluminescence*, ed. E. Von Dyke, pp. 173–92. Boca Raton: CRC Press.

Halliwell, B. & Gutteridge, J.M.C. (1984). Oxygen toxicity, oxygen radicals, transition metals and disease. *Biochemical Journal* **219**, 1–14.

Halliwell, B. & Gutteridge, J.M.C. (1985). *Free Radicals in Biology and Medicine*, p. 346. Oxford: Clarendon Press.

Hancock, J.T., Maly, F.-E. & Jones, O.T.G. (1989). Properties of the superoxide-generating oxidase of B-lymphocyte cell lines. Determination of Michaelis parameters. *Biochemical Journal* **262**, 373–5.

Harper, A.M., Chaplin, M.F. & Segal, A.W. (1985). Cytochrome b_{-245} from human neutrophils is a glycoprotein. *Biochemical Journal* **227**, 783–8.

Harper, A.M., Dunne, M.J. & Segal, A.W. (1984). Purification of cytochrome b_{-245} from human neutrophils. *Biochemical Journal* **219**, 519–27.

Hattori, H. (1961). Studies on the labile, stable NADH oxidase and peroxidase staining reactions in the isolated particles of horse granulocytes. *Nagoya Journal of Medical Science* **23**, 362–78.

Hayakawa, T., Suzuki, K., Suzuki, S., Andrews, P.C. & Babior, B.M. (1986). A possible role for protein phosphorylation in the activation of the respiratory burst in human neutrophils. Evidence from studies with cells from patients with chronic granulomatous disease. *Journal of Biological Chemistry* **261**, 9109–15.

Henderson, L.M., Chappell, J.B. & Jones, O.T.G. (1989). Superoxide generation is inhibited by phospholipase A_2 inhibitors. Role for phospholipase A_2 in the activation of the NADPH oxidase. *Biochemical Journal* **264**, 249–55.

Heyneman, R.A. & Vercauteren, R.E. (1984). Activation of a NADPH oxidase from horse polymorphonuclear leukocytes in a cell free system. *Journal of Leukocyte Biology* **36**, 751–9.

Heyworth, P.G. & Segal, A.W. (1986). Further evidence for the involvement of a phosphoprotein in the respiratory burst oxidase of human neutrophils. *Biochemical Journal* **239**, 723–31.

Heyworth, P.G., Shrimpton, C.F. & Segal, A.W. (1989). Localisation of the 47 kDa phosphoprotein involved in the respiratory-burst NADPH oxidase of phagocytic cells. *Biochemical Journal* **260**, 243–8.

Hidaka, H., Inagaki, M., Kawamoto, S. & Sasaki, Y. (1984).

Isoquinolinesulfonamides, novel and potent inhibitors of cyclic nucleotide dependent protein kinase and protein kinase C. *Biochemistry* **23**, 5036–41.

Holmes, B., Page, A.R. & Good, R.A. (1967). Studies of the metabolic activity of leukocytes from patients with a genetic abnormality of phagocytic function. *Journal of Clinical Investigation* **46**, 1422–32.

Hughes, V.A., Humphreys, J.M. & Edwards, S.W. (1987). Protein synthesis is activated in primed neutrophils: a possible role in inflammation. *Bioscience Reports* **7**, 881–9.

Huizinga, T.W.J., van der Schoot, C.E., Jost, C., Klassen, R., Kleijer, M., von dem Borne, A.E.G.Kr., Roos, D. & Tetteroo, P.A.T. (1988). The PI-linked receptor of FcR111 is released on stimulation of neutrophils. *Nature* **333**, 667–9.

Huizinga, T.W.J., van Kemenade, F., Koenderman, L., Dolman, K.M., von dem Borne, A.E.G.Kr., Tetteroo, P.A.T. & Roos, D. (1989). The 40-kDa Fcγ receptor (FcR11) on human neutrophils is essential for the IgG-induced respiratory burst and IgG-induced phagocytosis. *Journal of Immunology* **142**, 2365–9.

Humphreys, J.M., Hughes, V. & Edwards, S.W. (1989). Stimulation of protein synthesis in human neutrophils by γ-interferon. *Biochemical Pharmacology* **38**, 1241–6.

Iyer, G.Y.N., Islam, M.F. & Quastel, J.H. (1961). Biochemical aspects of phagocytosis. *Nature* **192**, 535–41.

Jack, R.M. & Fearon, D.T. (1988). Selective synthesis of mRNA and proteins by human peripheral blood neutrophils. *Journal of Immunology* **140**, 4286–93.

Jones, D.H., Anderson, D.C., Burr, B.L., Rudloff, H.E., Smith, C.W., Krater, S.S. & Schmalstieg, F.C. (1988). Quantitation of intracellular Mac-1 (CD11b/CD18) pools in human neutrophils. *Journal of Leukocyte Biology* **44**, 535–44.

Jones, D.H., Looney, R.J. & Anderson, C.L. (1985). Two distinct classes of IgG Fc receptors on human monocyte line (U937) determined by differences in binding of murine IgG subclasses at low ionic strength. *Journal of Immunology* **135**, 3348–53.

Jost, C.R., Huizinga, T.W.J., de Goede, R., Fransen, J.A.M., Tetteroo, P.A.T., Daha, M.R. & Ginsel, L.A. (1990). Intracellular localisation and de novo synthesis of FcRIII in human neutrophil granulocytes. *Blood* **75**, 144–51.

Kakinuma, K., Fukuhara, Y. & Kaneda, M. (1987). The respiratory burst oxidase of neutrophils. Separation of an FAD enzyme and its characterisation. *Journal of Biological Chemistry* **262**, 12316–22.

Kakinuma, K., Kaneda, M., Chiba, T. & Ohnishi, T. (1986). Electron spin resonance studies on a flavoprotein in neutrophil plasma membranes. Redox potentials of the flavin and its participation in NADPH oxidase. *Journal of Biological Chemistry* **261**, 9426–32.

Kawamoto, S. & Hidaka, H. (1984). 1-(5-isoquinolinesulfonyl)-2-

methylpiperazine (H-7) is a selective inhibitor of protein kinase C in rabbit platelets. *Biochemical and Biophysical Research Communications* **125**, 258–64.

Kettle, A.J. & Winterbourn, C.C. (1988). Superoxide modulates the activity of myeloperoxidase and optimises the production of hypochlorous acid. *Biochemical Journal* **252**, 529–36.

Klebanoff, S.J. (1968). Myeloperoxidase-halide-hydrogen peroxide antibacterial system. *Journal of Bacteriology* **95**, 2131–8.

Klebanoff, S.J. & Clark, R.A. (1978). *The Neutrophil: Function and Clinical Disorders*, p. 810. Amsterdam: North-Holland.

Kobayashi, S., Imajoh-Ohmi, S., Nakamura, M. & Kanegasaki, S. (1990). Occurrence of cytochrome b_{558} in B-cell lineage of human lymphocytes. *Blood* **75**, 458–61.

Koenderman, L., Tool, A., Roos, D. & Verhoeven, A.J. (1989). 1,2-Diacylglycerol accumulation in human neutrophils does not correlate with respiratory burst activation. *FEBS Letters* **243**, 399–403.

Kramer, I.M., Verhoeven, A.J., van der Bend, R.L., Weening, R.S. & Roos, D. (1988). Purified protein kinase C phosphorylates a 47 kDa protein in control neutrophil cytoplasts but not neutrophil cytoplasts from patients with the autosomal form of chronic granulomatous disease. *Journal of Biological Chemistry* **263**, 2352–7.

Kurzinger, K. & Springer, T.A. (1982). Purification and structural characteristics of LFA-1, a lymphocyte function-associated antigen, and Mac-1, a related macrophage differentiation antigen associated with the type three complement receptor. *Journal of Biological Chemistry* **257**, 12412–18.

Lackie, J.M. (1988). The behavioural repertoire of neutrophils requires multiple signal transduction pathways. *Journal of Cell Science* **89**, 449–52.

Lackie, J.M. & Lawrence, A.J. (1987). Signal response transduction in rabbit neutrophil leucocytes. The effects of exogenous phospholipase A_2 suggest that two pathways exist. *Biochemical Pharmacology* **36**, 1941–5.

Lehrer, R.I., Ganz, T., Selsted, M.E., Babior, B.M. & Curnutte, J.T. (1988). Neutrophils and host defence. *Annals of Internal Medicine* **109**, 127–42.

Leto, T.L., Lomax, K.J., Volpp, B.D., Nunoi, H., Sechler, J.M.G., Nauseef, W.M., Clark, R.A., Gallin, J.I. & Malech, H.L. (1990). Cloning of a 67-kD Neutrophil oxidase factor with similarity to a non-catalytic region of $p60^{c\text{-}src}$. *Science* **248**, 727–30.

Light, D.R., Walsh, C., O'Callaghan, A.M., Goetzl, E.J. & Tauber, A.I. (1981). Characteristics of the co-factor requirements for the superoxide-generating NADPH oxidase of human polymorphonuclear leukocytes. *Biochemistry* **20**, 1468–76.

Lindemann, A., Riedel, D., Oster, W., Meuer, S.C., Blohm, D., Mertelsmann, R. & Herrmann, F. (1988). Granulocyte/macrophage colony-stimulating factor induces interleukin-1 production by polymorphonuclear leukocytes. *Journal of Immunology* **140**, 837–9.

Lindemann, A., Riedel, D., Oster, W., Ziegler-Heitbrock, H.W.L., Mertelsmann, R. & Herrmann, F. (1989). Granulocyte-macrophage colony-stimulating factor induces cytokine secretion by human polymorphonuclear leukocytes. *Journal of Clinical Investigation* **83**, 1308–12.

Lomax, K.J., Leto, T.L., Nunoi, H., Gallin, J.I. & Malech, H.L. (1989). Recombinant 47-kilodalton cytosol factor restores NADPH oxidase in chronic granulomatous disease. *Science* **245**, 409–12.

Looney, R.J., Abraham, G.N. & Anderson, C.L. (1986). Human monocytes and U937 cells bear two distinct Fc receptors for IgG. *Journal of Immunology* **136**, 1641–7.

Lopez, A.F., Williamson, J., Gamble, J.R., Begley, C.G., Harlan, J.M., Klebanoff, S.J., Waltersdorph, A., Wong, G., Clark, S.C. & Vadas, M.A. (1986). Recombinant human granulocyte-macrophage colony-stimulating factor stimulates *in vitro* mature human neutrophil and eosinophil function, surface receptor expression, and survival. *Journal of Clinical Investigation* **78**, 1220–8.

Maly, F.E., Cross, A.R., Jones, O.T.G., Wolf-Vorbeck, Walker, C., Dahinden, C.A. & De Weck, A.L. (1988). The superoxide generating system of B cell lines. Structural homology with the phagocytic oxidase and triggering via surface Ig. *Journal of Immunology* **140**, 2334–9.

Maly, F.E., Nakamura, M., Gauchat, J.-F., Urwyler, A., Walker, C., Dahinden, C.A., Cross, A.R., Jones, O.T.G. & De Weck, A.L. (1989). Superoxide-dependent nitroblue tetrazolium reduction and expression of cytochrome b_{-245} components by human tonsillar B lymphocytes and B cell lines. *Journal of Immunology* **142**, 1260–7.

Maridonneau-Parini, I., Clerc, J. & Polla, B.S. (1988). Heat shock inhibits NADPH oxidase in human neutrophils. *Biochemical and Biophysical Research Communications* **154**, 179–86.

Maridonneau-Parini, I., Tringale, S.M. & Tauber, A.I. (1986). Identification of distinct activation pathways of the human neutrophil NADPH-oxidase. *Journal of Immunology* **137**, 2925–9.

Marino, J.A., Davis, A.H. & Spagnuolo, P.J. (1987). Fibronectin is stored but not synthesized in mature human peripheral blood granulocytes. *Biochemica et Biophysica Acta* **146**, 1132–8.

Markert, M., Glass, G.A. & Babior, B.M. (1985). Respiratory burst oxidase from human neutrophils: purification and some properties. *Proceedings of the National Academy of Sciences USA* **82**, 3144–8.

McCall, C.E., Bass, D.A., DeChatelet, L.R., Link, A.S. Jr. & Mann, M. (1979). *In vitro* responses of human neutrophils to N-formyl-methionyl-leucyl-phenylalanine: correlation with effects of acute bacterial infection. *Journal of Infectious Diseases* **140**, 277–86.

McPhail, L.C., Clayton, C.C. & Snyderman, R. (1984). The NADPH oxidase of human polymorphonuclear leucocytes. Evidence for regulation by multiple signals. *Journal of Biological Chemistry* **259**, 5768–75.

McPhail, L.C., Shirley, P.S., Clayton, C.C. & Snyderman, R. (1985). Activation of the respiratory burst enzyme from human neutrophils in a cell-free system: evidence for a soluble factor. *Journal of Clinical Investigation* **75**, 1735–9.

Miller, L.J., Bainton, D.F., Borregaard, N. & Springer, T.A. (1987). Stimulated mobilisation of monocyte Mac-1 and p150,95 adhesion proteins from an intracellular vesicular compartment to the cell surface. *Journal of Clinical Investigation* **80**, 535–44.

Nasmith, P.E., Mills, G.B. & Grinstein, S. (1989). Guanine nucleotides induce tyrosine phosphorylation and activation of the respiratory burst in neutrophils. *Biochemical Journal* **257**, 893–7.

Neer, E.J. & Clapham, D.E. (1988). Roles of G protein subunits in transmembrane signalling. *Nature* **333**, 129–34.

Newburger, P.E., Ezekowitz, A.B., Whitney, C., Wright, J. & Orkin, S.H. (1988). Induction of phagocyte cytochrome *b* heavy chain gene expression by interferon γ. *Proceedings of the National Academy of Sciences USA* **85**, 5215–19.

Nugent, J.H.A., Gratzer, W. & Segal, A.W. (1989). Identification of the haem-binding subunit of cytochrome b_{-245}. *Biochemical Journal* **264**, 921–4.

Nunoi, H., Rotrosen, D., Gallin, J.I. & Malech, H.L. (1988). Two forms of autosomal chronic granulomatous disease lack distinct neutrophil cytosol factors. *Science* **242**, 1298–301.

O'Shea, J.J., Brown, E.J., Seligman, B.E., Metcalf, J.A., Frank, M.M. & Gallin, J.I. (1984). Evidence for distinct intracellular pools of receptors for C3b and C3bi in human neutrophils. *Journal of Immunology* **134**, 2580–7.

Odajima, T. & Yamazaki, I. (1972). Myeloperoxidase of the leukocyte of normal blood. IV. Some physicochemical properties. *Biochemica et Biophysica Acta* **284**, 360–7.

Ohno, Y., Buescher, E.S., Roberts, R., Metcalf, J.A. & Gallin, J.I. (1986). Reevaluation of cytochrome b and flavin dinucleotide in neutrophils from patients with chronic granulomatous disease and description of a family with probable autosomal recessive inheritance of cytochrome b deficiency. *Blood* **67**, 1132–8.

Okajima, F., Katada, T. & Ui, M. (1985). Coupling of the guanine nucleotide regulatory protein to chemotactic peptide receptors in neutrophil membranes and its uncoupling by islet-activating protein. A possible role of the toxin substrate in Ca^{2+}-mobilising receptor-mediated signal transduction. *Journal of Biological Chemistry* **260**, 6761–8.

Okamura, N., Malawista, S.E., Roberts, R.L., Rosen, H., Ochs, H.D., Babior, B.M. & Curnette, J.T. (1988). Phosphorylation of the oxidase related 48K phosphoprotein family in the unusual autosomal cytochrome-negative and X-linked cytochrome-positive types of chronic granulomatous disease. *Blood* **72**, 811–16.

Parkos, C.A., Allen, R.A., Cochrane, C.G. & Jesaitis, A.J. (1987). Cytochrome *b* from human granulocyte plasma membrane is comprised of two polypeptides with relative molecular weights of 91,000 and 22,000. *Journal of Clinical Investigation* **80**, 732–42.

Parkos, C.A., Dinauer, M.C., Walker, L.E., Allen, R.A., Jesaitis, A.J. & Orkin, S.H. (1988). Primary structure and unique expression of the 22-kilodalton light chain of human neutrophil cytochrome *b*. *Proceedings of the National Academy of Sciences USA* **85**, 3319–23.

Perussia, B., Dayton, E.T., Lazarus, R., Fanning, V. & Trinchieri, G. (1983). Immune interferon induces the receptor for monomeric IgGl on human monocyte and myeloid cells. *Journal of Experimental Medicine* **158**, 1092–113.

Petrequin, P.R., Todd, R.F.III, Devall, L.J., Boxer, L.A. & Curnutte, J.T.III. (1987). Association between gelatinase release and increased plasma membrane expression of the Mol glycoprotein. *Blood* **69**, 605–10.

Petroni, K.C., Shen, L. & Guyre, P.M. (1988). Modulation of human polymorphonuclear leukocyte IgG Fc receptors and Fc receptor-mediated functions by INF-γ and glucocorticoids. *Journal of Immunology* **140**, 3467–72.

Philips, M.R., Buyon, J.P., Winchester, R., Weissmann, G. & Abramson, S.B. (1988). Up-regulation of the iC3b receptor (CR3) is neither necessary nor sufficient to promote neutrophil aggregation. *Journal of Clinical Investigation* **82**, 495–501.

Pou, S., Cohen, M.S., Britigan, B.E. & Rosen, G.M. (1989). Spin-trapping and human neutrophils. Limits of detection of hydroxyl radical. *Journal of Biological Chemistry* **264**, 12299–302.

Pozzan, T., Lew, D.., Wollheim, C.B. & Tsien, R.Y. (1982). Is cytosolic ionised calcium regulating neutrophil activation? *Science* **221**, 1413–15.

Reinhold, S.L., Prescott, S.M., Zimmerman, G.A. & McIntyre, T.M. (1990). Activation of human neutrophil phospholipase D by three separable mechanisms. *FASEB Journal* **4**, 208–14.

Ross, G.D. & Medof, M.E. (1985). Membrane complement receptors specific for bound fragments of C3. *Advances in Immunology* **37**, 217–67.

Rossi, F. (1986). The O_2^- forming NADPH oxidase of the phagocytes: nature, mechanisms of activation and function. *Biochemica et Biophysica Acta* **853**, 65–89.

Royer-Pokora, B., Kunkel, L.M., Monaco, A.P., Goff, S.C., Newburger, P.E., Baehner, R.L., Cole, F.S., Curnutte, J.T. & Orkin, S.H. (1986). Cloning the gene for an inherited human disorder – chronic granulomatous disease – on the basis of its chromosomal location. *Nature* **322**, 32–8.

Sadler, K.L. & Badwey, J.A. (1988). Second messengers involved in superoxide production by neutrophils: function and metabolism. In *Phagocyte Defects II. Hematology/Oncology Clinics of North America*, ed. J.T. Curnutte, pp. 185–200. Philadelphia: Saunders.

Sanchez-Madrid, F., Nagy, J.A., Robbins, E., Simon, P. & Springer, T.A. (1983). A human leukocyte differentiation antigen family with distinct α subunits and a common β subunit: the lymphocyte function-associated antigen (LFA-1), the C3bi complement receptor (OKMI/Mac-1) and the p150,95 molecule. *Journal of Experimental Medicine* **158**, 1785–803.

Sbarra, A.J. & Karnovsky, M.L. (1959). The biochemical basis of phagocytosis. 1. Metabolic changes during the ingestion of particles by polymorphonuclear leukocytes. *Journal of Biological Chemistry* **234**, 355–62.

Segal, A.W. (1987). Absence of both cytochrome b_{-245} subunits from neutrophils in X-linked chronic granulomatous disease. *Nature* **326**, 88–91.

Segal, A.W. (1988). Cytochrome b_{-245} and its involvement in the molecular pathology of chronic granulomatous disease. In *Phagocytic Defects II. Hematology/Oncology Clinics of North America*, ed. J.T. Curnutte, pp. 213–24. Philadelphia: Saunders.

Segal, A.W. (1989*a*). The electron transport chain of the microbicidal oxidase of phagocytic cells and its involvement in the molecular pathology of chronic granulomatous disease. *Biochemical Society Transactions* **17**, 427–34.

Segal, A.W. (1989*b*). The electron transport chain of the microbicidal oxidase of phagocytic cells and its involvement in the molecular pathology of chronic granulomatous disease. *Journal of Clinical Investigation* **83**, 1785–93.

Segal, A.W., Cross, A.R., Garcia, R.C., Borregaard, N., Valerius, N.H., Soothill, J.F. & Jones, O.T.G. (1983). Absence of cytochrome b_{-245} in chronic granulomatous disease. A multicentre European evaluation of its incidence and relevence. *New England Journal of Medicine* **308**, 245–51.

Segal, A.W., Garcia, R., Goldstone, A.H., Cross, A.R. & Jones, O.T.G. (1981). Cytochrome b_{-245} of neutrophils is also present in human monocytes, macrophages and eosinophils. *Biochemical Journal* **196**, 363–7.

Segal, A.W., Heyworth, P.G., Cockcroft, S. & Barrowman, M.M. (1985). Stimulated neutrophils from patients with autosomal recessive chronic granulomatous disease fail to phosphorylate a Mr-44,000 protein. *Nature* **316**, 547–49.

Segal, A.W. & Jones, O.T.G. (1978). Novel cytochrome *b* system in phagocytic vacuoles of human granulocytes. *Nature* **276**, 515–17.

Segal, A.W., Jones, O.T.G., Webster, D. & Allison, A.C. (1978). Absence of a newly described cytochrome b from neutrophils of patients with chronic granulomatous disease. *Lancet* **ii**, 446–9.

Selvaraj, P., Rosse, W.F., Silber, R. & Springer, T.A. (1988). The major Fc receptor in blood has a phosphoinositol anchor and is deficient in paroxysmal nocturnal haemoglobinuria. *Nature* **333**, 565–7.

Selvaraj, R.J. & Sbarra, A.J. (1966). Relationship of glycolytic and oxidative metabolism to particle entry and destruction in phagocytosing cells. *Nature* **211**, 1272–6.

Shaw, S. (1987). Characterisation of human leukocyte differentiation antigens. *Immunology Today* **8**, 1–13.

Shirafuji, N., Matsuda, S., Ogura, H., Tani, K., Kodo, H., Ozawa, K., Nagata, S., Asano, S. & Takaku, F. (1990). Granulocyte colony-stimulating factor stimulates human mature neutrophilic granulocyte to produce interferon-α. *Blood* **75**, 17–19.

Simmons, D. & Seed, B. (1988). The Fcγ receptor of natural killer cells is a phospholipid-linked membrane protein. *Nature* **333**, 568–70.

Smith, C.D., Cox, C.C. & Snyderman, R.(1986). Receptor-coupled activation of phosphoinositide-specific phopspholipase C by an N protein. *Science* **232**, 97–100.

Spitznagel, J.K. (1984). Nonoxidative antimicrobial systems in leucocytes. In *Regulation of Leukocyte Function (Contemporary Topics in Immunology)*, vol. 14, ed. R. Snyderman, pp. 234–83. New York: Plenum.

Stevenson, K.B., Nauseef, W.M. & Clark, R.A. (1987). The neutrophil glycoprotein Mo1 is an integral membrane protein of plasma membranes and specific granules. *Journal of Immunology* **139**, 3759–63.

Tamaoki, T., Nomoto, H., Takahashi, I., Kato, Y., Morimoto, M. & Tomita, F. (1986). Staurosporine, a potent inhibitor of phospholipid/Ca^{++} dependent protein kinase. *Biochemical and Biophysical Research Communications* **135**, 397–402.

Tauber, A.I. (1987). Protein kinase c and the activation of the human neutrophil NADPH oxidase. *Blood* **69**, 711–20.

Tauber, A.I., Borregaard, N., Simons, E. & Wright, J. (1983). Chronic granulomatous disease: a syndrome of phagocyte oxidase deficiencies. *Medicine* **62**, 286–309.

Todd, R.F.III & Freyer, D.R. (1988). The CD11/CD18 leukocyte glycoprotein deficiency. In *Phagocytic Defects I. Hematology/Oncology Clinics of North America*, pp. 13–32.

Todd, R.F.III, Arnaout, M.A., Rosin, R.E., Crowley, C.A., Peters, W.A. & Babior, B.M. (1984). Subcellular localisation of the large subunit of Mo1 (Mo1-α; formerly gp110), a surface glycoprotein associated with neutrophil adhesion. *Journal of Clinical Investigation* **74**, 1280–90.

Van Epps, D.E. & Garcia, M.L. (1980). Enhancement of neutrophil function as a result of prior exposure to chemotactic factor. *Journal of Clinical Investigation* **56**, 167–75.

Vedder, N.B. & Harlan, J.M. (1988). Increased surface expression of

CD11b/CD18 (Mac-1) is not required for stimulated neutrophil adherence to cultured endothelium. *Journal of Clinical Investigation* **81**, 676–82.

Volpp, B.D., Nauseef, W.M. & Clark, R.A. (1988). Two cytosolic neutrophil oxidase components absent in autosomal chronic granulomatous disease. *Science* **242**, 1295–7.

Volpp, B.D., Nauseef, W.M., Donelson, J.E., Moser, D.R. & Clark, R.A. (1989). Cloning of the cDNA and functional expression of the 47-kilodalton cytosolic component of human neutrophil respiratory burst oxidase. *Proceedings of the National Academy of Sciences USA* **86**, 7195–9.

Weisbart, R.H. (1989). Colony-stimulating factors and host defence. *Annals of Internal Medicine* **110**, 297–303.

Weisbart, R.H., Golde, D.W. & Gasson, J.C. (1986). Biosynthetic human GM-CSF modulates the number and affinity of neutrophil f-Met–Leu–Phe receptors. *Journal of Immunology* **137**, 3584–7.

Winterbourn, C.C. (1986). Myeloperoxidase as an effective inhibitor of hydroxyl radical production. Implications for the oxidative reactions of neutrophils. *Journal of Clinical Investigation* **78**, 545–50.

Winterbourn, C.C., Garcia, R.C. & Segal, A.W. (1985). Production of the superoxide adduct of myeloperoxidase (compound III) by stimulated human neutrophils and its reactivity with hydrogen peroxide and chloride. *Biochemical Journal* **228**, 583–92.

Wolf, M. & Baggiolini, M. (1988). The protein kinase inhibitor staurosporine, like phorbol esters, induces the association of protein kinase C with membranes. *Biochemical and Biophysical Research Communications* **154**, 1273–9.

Wood, P.M. (1987). The two redox potentials for oxygen reduction to superoxide. *Trends in Biochemical Sciences* **12**, 250–1.

Wright, S.D., Rao, P.E., Van Voorhis, W.C., Craigmyle, L.S., Iida, K., Talle, M.A., Westberg, E.F., Goldstein, G. & Silverstein, S.C. (1983). Identification of the C3bi receptor on human leukocytes and macrophages using monoclonal antibodies. *Proceedings of the National Academy of Sciences USA* **80**, 5699–703.

Wymann, M.P., von Tscharner, V., Deranleau, D.A. & Baggiolini, M. (1987). The onset of the respiratory burst in human neutrophils. Real-time studies of H_2O_2 formation reveal a rapid agonist-induced transduction process. *Journal of Biological Chemistry* **262**, 12048–53.

Yamagata, N., Kobayashi, K., Kasama, T., Fukushma, T., Tabata, M., Yoneya, I., Shilama, Y., Kaya, S., Hashimoto, M., Yoshida, K., Sekine, F., Negishi, F., Ide, H., Mori, Y. & Takahashi, T. (1988). Multiple cytokine activities and loss of interleukin 2 inhibitor in synovial fluids of patients with rheumatoid arthritis. *Journal of Rheumatology* **15**, 1623–7.

Yamaguchi, T., Hayakawa, T., Kaneda, M., Kakinuma, K. & Yoshikawa, A. (1989). Purification and some properties of the small subunit of cytochrome b_{558} from human neutrophils. *Journal of Biological Chemistry* **264**, 112–18.

Yamazaki, M., Ikenami, M. & Sugiyama, T. (1989). Cytotoxin from polymorphonuclear leukocytes and inflammatory ascitic fluids. *British Journal of Cancer* **59**, 353–5.

Yea, C.M., Cross, A.R. & Jones, O.T.G. (1990). Purification and some properties of the 45 kDa diphenylene iodonium-binding flavoprotein of neutrophil NADPH oxidase. *Biochemical Journal* **265**, 95–100.

MICHAEL J. SHATTOCK, HIROSHI MATSUURA and DAVID J. HEARSE

Reperfusion arrhythmias: role of oxidant stress

Introduction

Reperfusion of the ischaemic myocardium

The occlusion of a coronary artery leads to the development of myocardial ischaemia and initiates the process of infarction. It is clear that the readmission of blood to severely ischaemic tissue is essential if that tissue is ultimately to survive. Clinically, reperfusion can be achieved within hours of the onset of an ischaemic episode (using angioplasty and thrombolytic techniques), following prolonged ischaemia (using coronary artery bypass grafts) or may occur spontaneously following coronary artery spasm. The recent development of safe angioplasty techniques and effective thrombolytic drugs has led to an increased interest in the consequences of reperfusion. Although clinically such interventions can preserve myocardial function and improve prognosis (Serruys *et al.*, 1986), there is a growing body of evidence from animal models to suggest that reperfusion can, in itself, promote a variety of undesirable effects such as arrhythmias (Manning & Hearse, 1984), myocardial stunning (Braunwald & Kloner, 1982), and leucocyte infiltration and vasoconstriction (Engler *et al.*, 1986).

The concept that reperfusion of reversibly damaged but viable cells can promote lethal injury in that tissue is, however, controversial (Hearse, 1989; 1991; Opie, 1989). Although there is considerable experimental evidence implying that reperfusion may *accelerate* the expression of injury in cells that are ultimately destined to die, conclusive evidence that reperfusion itself can kill viable cells does not yet exist. The acceleration of tissue injury by reperfusion may, however, alter the processes underlying cell death and may therefore lead to differences in scar formation, leucocyte infiltration and susceptibility to aneurism. There is considerable evidence in support of the existence of this type of reperfusion-induced injury.

The precise mechanisms underlying the adverse effects of reperfusion have yet to be definitively characterised. While it is unlikely that a single factor underlies the many facets of reperfusion injury, the readmission of oxygen appears to play a central role. Cellular damage following reperfusion is proportional not only to the period of ischaemia but also to the pO_2 of the reperfusing medium (Hearse, Humphrey & Chain, 1973; Hearse *et al.*, 1975; Hearse, Humphrey & Bullock, 1978). In these studies of ischaemia and hypoxia, cellular damage during reperfusion or reoxygenation was shown to be far greater than that which would have occurred if the ischaemia or hypoxia had been maintained. The central role of the readmission of oxygen in the genesis of reperfusion injury has led to the suggestion that oxygen-derived free radicals and consequent oxidant stress may contribute to reperfusion injury.

Sources of free-radical-induced oxidant stress in the heart

There are a number of potential sources of reactive oxygen species in the heart both under normal physiological conditions and during ischaemia, hypoxia, reperfusion or reoxygenation. For example, under normal aerobic conditions approximately 2% of molecular oxygen in the mitochondria is reduced to water via a series of univalent steps involving the generation of reactive oxygen species (superoxide anion, hydrogen peroxide and the hydroxyl radical) (Boveris & Chance, 1973). Free radicals are also produced by the auto-oxidation of catecholamines and the breakdown of myoglobin and haemoglobin (Caughey & Watkins, 1985). The arachidonic acid pathway (Halliwell & Gutteridge, 1985), invading leucocytes (Romson *et al.*, 1983) and many oxidase enzymes (such as NADPH oxidase and xanthine oxidase) are also important potential sources of radicals in the heart. The xanthine oxidase pathway has recently received considerable attention as a possible source of superoxide radicals in ischaemia and reperfusion (Chambers *et al.*, 1985). There is, however, some controversy as to whether xanthine oxidase is present in human myocardium, whether it is only present in the endothelium, or whether it is entirely absent (see Hearse, 1989).

Under normal conditions, myocardial cells are equipped with a variety of cellular defences against reactive oxygen species and oxidant stress. These include superoxide dismutase, glutathione peroxidase, catalase, α-tocopherol and ascorbic acid. Thus, under normal conditions, the cell can adequately deal with the endogenous production of free radicals and oxidant stress. However, during ischaemia and reperfusion, where cellular defences may be altered or overwhelmed, free radicals may play an important role in the processes of cellular injury.

Oxidant stress during ischaemia and reperfusion

A number of cellular and metabolic changes occur during ischaemia which serve both to lower the ability of the cell to defend itself against oxidant stress and to promote radical-generating reactions. Ischaemia impairs endogenous anti-oxidant activity, such as the activity of SOD (Hammond & Hess, 1985), and thus renders the cell more susceptible to subsequent oxidant stress. In addition, there is an ischaemia-induced breakdown of adenine nucleotides, the subsequent accumulation of xanthine and hypoxanthine, and the conversion of xanthine dehydrogenase to xanthine oxidase (Bindoli *et al.*, 1988). The presence of xanthine and xanthine oxidase, the activation of catecholamine and arachidonic acid metabolism, and the release of chemotactic stimuli and the subsequent infiltration of leucocytes, all combine to create a situation which will promote oxidant stress. During ischaemia the low pO_2 will limit this oxidant stress and it is thus during reperfusion, with the rapid resupply of oxygen, that these ischaemia-induced changes may profoundly stimulate free-radical production and oxidant stress. In this regard, Hess & Manson (1984) have suggested that ischaemia acts to 'prime' the myocardium for free-radical production during the early moments of reperfusion.

Reperfusion-induced arrhythmias: evidence implicating free-radical-induced oxidant stress

Under a number of experimental conditions, the most profound and immediate consequence of reperfusion after a period of regional ischaemia is the induction of cardiac arrhythmias (Manning & Hearse, 1984). Oxidant stress has been implicated in the initiation of these arrhythmias and the evidence for this falls into three broad categories. Firstly, there are a number of studies showing that there is a burst of free-radical production in early reperfusion; secondly, anti-free-radical interventions have been shown to be anti-arrhythmic; and thirdly, free-radical generators can cause arrhythmias. The evidence implicating oxidant stress in reperfusion arrhythmias will be briefly considered in the following sections. It is, however, important to stress at this point that oxidant stress is by no means the only possible cause of reperfusion arrhythmias (Hearse & Tosaki, 1988). Yamada, Hearse & Curtis (1990) have recently demonstrated that arrhythmias can still be elicited when reperfusion is carried out with anoxic medium (pO_2 less than detectable levels). In this study, the subsequent readmission of oxygen did induce additional arrhythmias implying that although oxidant stress can induce arrhythmias, it is not a prerequisite. Other factors that have been suggested to contribute to arrhythmogenesis during early reperfusion include inhomogeneity of extracellular potassium gradients and potassium washout (Curtis, 1989), increases in α-receptor number (Sheridan *et al.*, 1980; Corr *et al.*, 1981), changes in metabolic status (Dennis *et al.*, 1983) and the intracellular accumulation of lysophosphatides (Corr & Sobel, 1982).

Production of oxygen-derived radicals in early reperfusion

Using Electron Spin Resonance techniques, a number of authors have demonstrated that there is a burst of free-radical production during the early moments of reperfusion (Misra *et al.*, 1984; Garlick *et al.*, 1987; Zweier, Flaherty & Weisfeldt, 1987; Bolli *et al.*, 1988). This transient period of free-radical production occurs over a similar time-course to that of reperfusion-induced arrhythmias (Manning & Hearse, 1984). Garlick *et al.* (1987) demonstrated that the burst of free-radical production is dependent on the readmission of O_2 rather than the readmission of flow. In addition, this burst could be attenuated by anti-oxidants such as superoxide dismutase and α-phenyl *N-tert*-butyl nitrone (Kramer *et al.*, 1987; Bolli *et al.*, 1988; Hearse & Tosaki, 1988).

Effects of anti-free-radical interventions

The observation that anti-free-radical interventions can exert profound anti-arrhythmic effects provides circumstantial evidence implicating oxidant stress in the induction of reperfusion-induced arrhythmias. Bernier, Hearse & Manning (1986), for example, demonstrated that superoxide dismutase, catalase, mannitol, glutathione and desferrioxamine can all reduce the incidence of reperfusion-induced ventricular fibrillation. These agents have widely different chemical structures but all share a common ability to either inhibit the formation of radicals or scavenge radicals after they are formed.

Effects of oxidant stress and free-radical-generating systems

Oxidant stress and free radicals, in the absence of ischaemia, hypoxia, reperfusion or reoxygenation, have been shown to induce automaticity in isolated ventricular muscle (Pallandi, Perry & Campbell, 1987; Matsuura & Shattock 1989*a*), single ventricular myocytes (Barrington, Meier & Weglicki, 1988; Matsuura & Shattock, 1989*b*) and isolated perfused hearts (Hearse, Kusama & Bernier, 1989). The cellular basis of oxidant-stress-induced arrhythmias has yet to be fully characterised but a number of studies have indicated that free-radical-induced damage may involve the induction of intracellular calcium overload (Borgers, Ver Donck & Vandeplassche, 1987; Burton *et al.*, 1988; Vandeplassche *et al.*, 1990) and there is a growing body of evidence indicating that oxidant stress can damage membrane pumps (Kim & Akera, 1987; Matsuoka *et al.*, 1987; Wang *et al.*, 1988; Kaneko, Beamish & Dhalla, 1989*a*), ion exchangers (Reeves, Bailey & Hale, 1986; Kato, Kako & Nagano, 1988; Shi, Davidson & Tibbetts, 1989) and ion channels (Kaneko *et al.*, 1989*b*).

Oxidant stress and the membrane potential

Few studies have characterised the influence of free radicals on basic cellular electrophysiology. Pallandi *et al.* (1987) demonstrated a free-radical-induced reduction in action potential amplitude, resting potential and upstroke velocity in guinea pig ventricular strips with no associated change in action potential duration. In these studies, 50% of preparations showed automaticity. A biphasic change in action potential duration has since been reported in canine (Barrington *et al.*, 1988) and frog (Tarr & Valenzeno, 1989) myocytes exposed to extracellularly applied free-radical-generating systems. Barrington *et al.* (1988) also reported failure of the action potential to repolarise and, in some cells, the induction of automaticity.

We have recently demonstrated that, even in the absence of ischaemia and reperfusion, oxidant stress generated by the photoactivation of rose bengal can induce rapid electrophysiological changes and arrhythmias (Hearse *et al.*, 1989). We have therefore undertaken a number of studies designed to investigate the cellular basis of rose-bengal-induced arrhythmias.

Studies in isolated muscles

On exposure to rose bengal (10 nmoles/l) and photoactivation, rat ventricular muscle showed a transient positive inotropy, after-contractions, after-depolarisations, prolongation of the action potential and the eventual development of contracture (Matsuura & Shattock, 1989*a*). These observations appear consistent with the possibility that free radicals induce cellular calcium overload (Borgers *et al.*, 1987; Burton *et al.*, 1988; Vandeplassche *et al.*, 1990).

The role of the sarcoplasmic reticulum (SR)

The after-contractions and after-depolarisations induced by oxidant stress in isolated papillary muscles were shown to be abolished with caffeine (10 mmoles/l) (Matsuura & Shattock 1989*a*). This suggests a possible role for oscillatory calcium release from the SR in these effects. The development of contracture, however, was not prevented by caffeine. This suggests that although the SR is involved in the expression of the calcium overload it is not necessarily the cause of the calcium overload *per se*. Further circumstantial evidence for the importance of the SR in the expression of these effects is that the potency of rose bengal in different species appears to be directly related to the extent to which SR calcium release contributes to excitation–contraction coupling (Bers, 1985). For example, rabbit ventricle shows similar responses to rat ventricle only at concentrations of rose bengal in excess of 100 nmoles/l and frog ventricle shows no after-contractions or after-depolarisations even at concentrations greater than 10 μmoles/l.

Although these results suggest that the SR may be important in the expression of free-radical-induced injury, direct free-radical-induced SR damage is not a requirement for oscillatory calcium release. There is evidence to suggest that simply overloading the cell with calcium is sufficient to induce spontaneous oscillatory SR calcium release (Kort & Lakatta, 1984).

The possibility that oxidant stress can directly influence SR function has, however, been recently investigated (Cumming *et al.*, 1989). In these studies, isolated SR calcium release channels were reconstituted into planar lipid bilayers and oxidant stress was shown to increase the open probability of the SR calcium release channel. Sub-conductance states, which are rarely seen in the intact channel, were also observed during oxidant stress. These results demonstrate that oxidant stress may indeed have a direct effect on SR calcium release and this may contribute to the arrhythmogenic and toxic effects of free radicals.

Oxidant stress and ionic currents

The calcium inward current

The observation that rose bengal-induced contractures still occur in caffeine-pretreated muscles (Matsuura & Shattock, 1989*a*), would suggest that in addition to influencing SR function, as described above, oxidant stress may influence other pathways responsible for cellular calcium homeostasis. Coetzee & Opie (1988) have suggested that the cellular calcium overload induced by free radicals is mediated by an increase in current through the voltage-activated calcium channel. However, we have recently investigated the influence of oxidant stress, induced by the photoactivation of rose bengal on the calcium inward current (i_{Ca}). Fig. 1 shows calcium currents recorded under control conditions and after 5 min of rose bengal (50 nmoles/l) photoactivation.

(*a*)

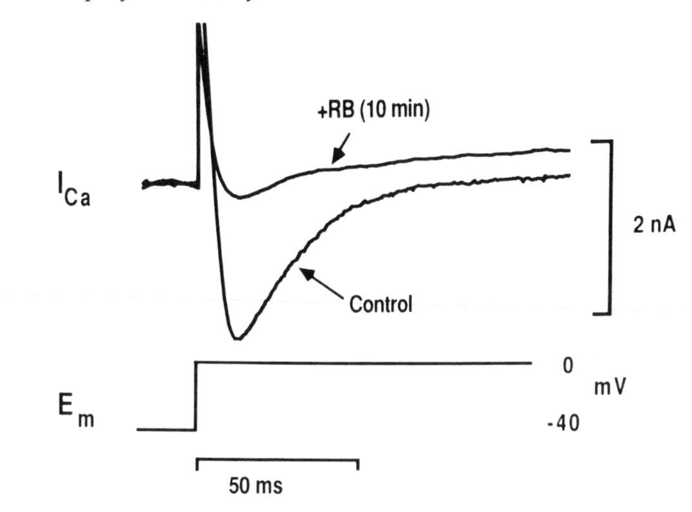

(*b*)

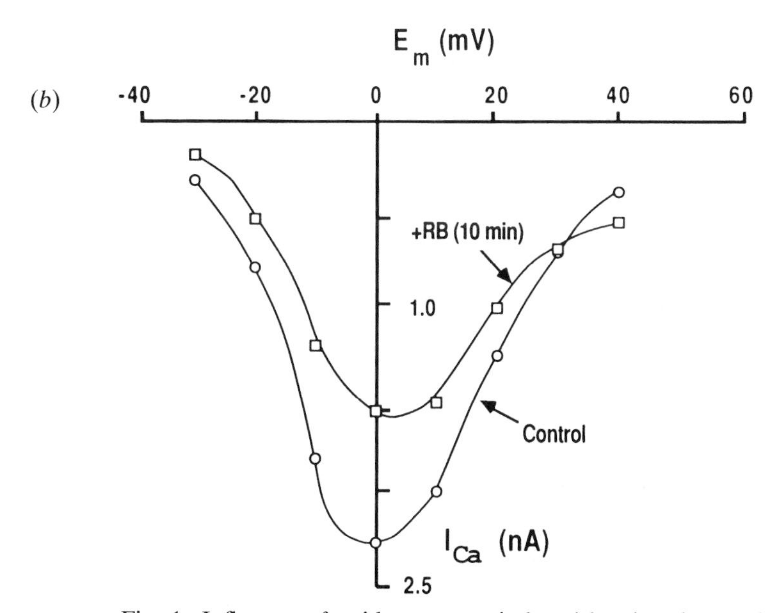

Fig. 1. Influence of oxidant stress, induced by the photoactivation of rose bengal (RB), on calcium currents recorded in rabbit ventricular myocytes. (*a*) Current records during a 500 ms clamp step to 0 mV. In addition to depressing i_{Ca}, RB also induced an inward shift in the background conductance and, in this example, the currents were offset so as to be superimposed at the start of the current trace. (*b*) Current–voltage relationship. To remove the effect of RB on the background conductance, peak current was measured with respect to current level at end of 500 ms clamp step. Holding potential = −40 mV.

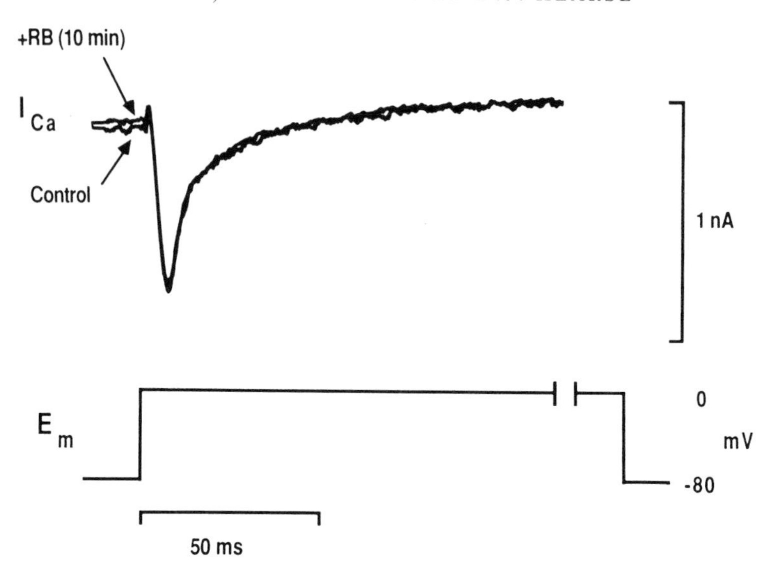

Fig. 2. Influence of oxidant stress, induced by the photoactivation of rose bengal (RB), on calcium currents recorded in rabbit ventricular myocytes following the removal of the transmembrane sodium gradient. Leak subtraction was performed by subtracting a scaled current record obtained during a small hyperpolarising clamp step. The composition of the Tyrode solution was (in mM) TEA Cl 140, CsCl 4, $CaCl_2$ 1.8, $MgCl_2$ 0.5, glucose 5.5, HEPES 5, (pH 7.4) and the pipette solution (in mM) $CsCl_2$ 135, $MgCl_2$ 1, ATP(Mg) 5, EGTA 10, glucose 10, HEPES 10 (pH 7.1). Holding potential = -80 mV, temp = 35 °C.

In rabbit myocytes, exposed to photoactivated rose bengal, peak calcium current was decreased and the slow time constant of i_{Ca} decay was, in fact, reduced by rose bengal from 45 to 25 ms. It therefore seems unlikely that the rose-bengal-induced cellular calcium overload is mediated via calcium influx through voltage-dependent channels.

Some insight into the possible route of calcium entry was gained when the above experiment was repeated in cells in which the transmembrane sodium gradient had been abolished. Intracellular sodium was reduced to zero (nominally) by dialysing the cell with a sodium-free solution in the patch pipette (potassium was also replaced with caesium) and extracellular sodium was then replaced with tetraethylammonium. Under these conditions, calcium currents can be recorded from a holding potential of -80mV and these currents run down slowly and are uncontaminated by other currents.

Fig. 2 shows calcium currents recorded using this approach under control conditions and after 10 min exposure to photoactivated rose bengal. In these experiments, in the absence of a transsarcolemmal sodium gradient, the calcium current was unaffected by exposure to photoactivated rose bengal. This observation suggests that oxidant stress does not directly affect the calcium channel. In addition, the cells exposed to rose bengal showed no evidence of calcium overload (unlike cells with a normal transmembrane sodium gradient which, after 5 min exposure, showed spontaneous oscillatory contractile waves and, after 10 min, complete contracture). These preliminary data suggest (i) that reactive oxygen intermediates do not directly affect the calcium channel; (ii) that the reduction of i_{Ca} in cells with a normal sodium gradient may be due to a Ca_i-induced inactivation of the calcium channel (Lee, Marban & Tsien, 1985); and (iii) that the sodium gradient is necessary to induce cellular calcium overload following exposure to reactive oxygen intermediates. This final observation implies that the Na/Ca exchange may be an important pathway in oxidant-stress-induced cellular calcium overload.

The Na-pump current

We have recently measured Na-pump current (i_p) in myocytes exposed to oxidant stress using the whole-cell voltage clamp technique (Shattock & Matsuura, 1990). In these experiments, pipette and bathing solutions were designed to block all voltage-gated channels and the Na/Ca exchange mechanism. The sodium concentration of the pipette solution was raised to 30 mmoles/l to maximally activate the Na-pump and cells were dialysed via wide-tipped electrodes (2–3 MΩ). Outward i_p was recorded during a ramp-pulse protocol and was identified by subtracting the current recorded in the absence of external potassium from that in its presence (5.4 mM).

Fig. 3*a* shows the current–voltage relationship of i_p under control conditions and during oxidant stress induced by the photoactivation of rose bengal. After 5 min exposure to rose bengal, i_p was reduced to 60% of control at 0 mV and to 70% of control at −75 mV. Fig. 3*b* shows currents recorded in the absence of extracellular potassium. Under these conditions, no active ionic currents remain (all channels, pumps and exchangers are blocked) and the remaining current reflects the passive resistance of the membrane. In the experiment shown in Fig. 3*b*, this was found to be approximately 2.5 GΩ. It is important to note that oxidant stress did *not* cause non-specific membrane damage and changes in passive membrane resistance.

Studies of Na/K-ATPase activity in the purified enzyme (Matsuoka *et al.*, 1987), in membrane vesicles (Wang *et al.*, 1988) and in isolated hearts

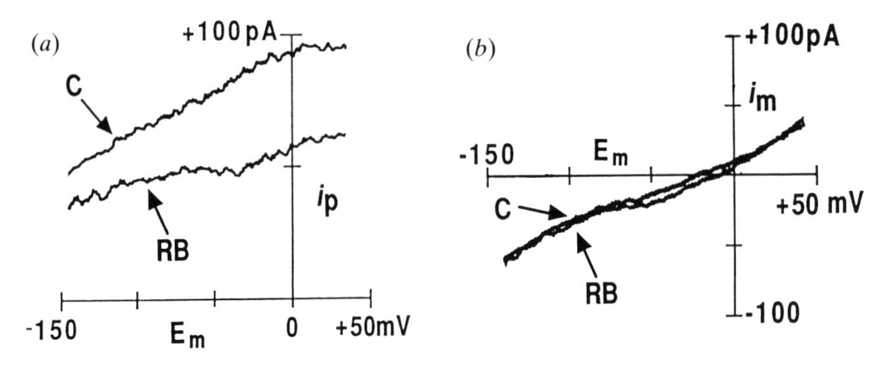

Fig. 3. Influence of oxidant stress, induced by the photoactivation of rose bengal, on Na-pump current. Pump current was recorded using the whole-cell voltage clamp technique and a wide-tipped (2–3 MΩ) suction electrodes. Bathing and pipette solutions were formulated to limit active channel currents and Na/Ca exchange and pump current was recorded during a ramp-pulse protocol (+50 to −150 mV) in the presence (*a*) and absence (*b*) of external potassium. The currents shown in panel *a* therefore show the influence of oxidant stress on Na-pump current while those in panel *b* show the influence of oxidant stress on passive membrane resistance. The composition of the Tyrode solution was (in mM) NaCl 140, KCl 5, CsCl 2, $BaCl_2$ 1, $NiCl_2$ 2, $MgCl_2$ 1, glucose 5.5, HEPES 5, (pH 7.4) and the pipette solution (in mM) Cs aspartate 100, NaCl 30, TEA Cl 20, $MgCl_2$ 2, ATP(Mg) 5, Creatine phosphate 5, EGTA 5, glucose 10, HEPES 10 (pH 7.2).

(Kim & Akera, 1987) have also demonstrated that the activity of the Na-pump is extremely sensitive to oxidant stress and free-radical-generating systems. In addition, Kim & Akera (1987) showed that anti-free-radical interventions could prevent ischaemia- and reperfusion-induced damage to the Na-pump. The inhibition of Na/K-ATPase activity by oxidant stress would lead to a rise in intracellular sodium and would favour an enhanced uptake of calcium via Na/Ca exchange.

The Na/Ca exchange

The observation that the cellular calcium overload, induced by oxidant stress, is dependent on the existance of a normal transsarcolemmal sodium gradient implies that the Na/Ca exchange mechanism may play a role in this process. The inhibition of the Na-pump by oxidant stress will tend to raise intracellular sodium and will promote a cellular calcium overload via the Na/Ca exchange. In addition, there is considerable evidence that calcium uptake via this pathway may be directly enhanced by a direct effect of free radicals on the Na/Ca exchange mechanism itself. In 1986, Reeves *et al.* published an important paper showing that free radical generating systems could stimulate, or 'disinhibit', the Na/Ca exchange mechanism. Kato *et al.* (1988) have also recently demonstrated that the Na/Ca exchange mechanism is stimulated by free radicals. There is no evidence to indicate whether the stoichiometry of the exchanger is altered by free-radical stimulation or to suggest that free-radical stimulation favours exchange primarily in the direction of calcium uptake. However, the combination of a prolongation of the action potential, a rise in intracellular sodium (due to Na-pump inhibition) and the stimulation of the Na/Ca exchange mechanism is likely to create a situation in which calcium influx via this pathway is greatly enhanced.

The transient inward current

In rabbit ventricular myocytes oxidant stress induced by the photoactivation of rose bengal activates an arrhythmogenic oscillatory transient inward current (i_{TI}) (Matsuura & Shattock, 1989*b*; 1991). Fig. 4 shows currents recorded under control conditions and after 5 min exposure to 50 nmoles/l rose bengal.

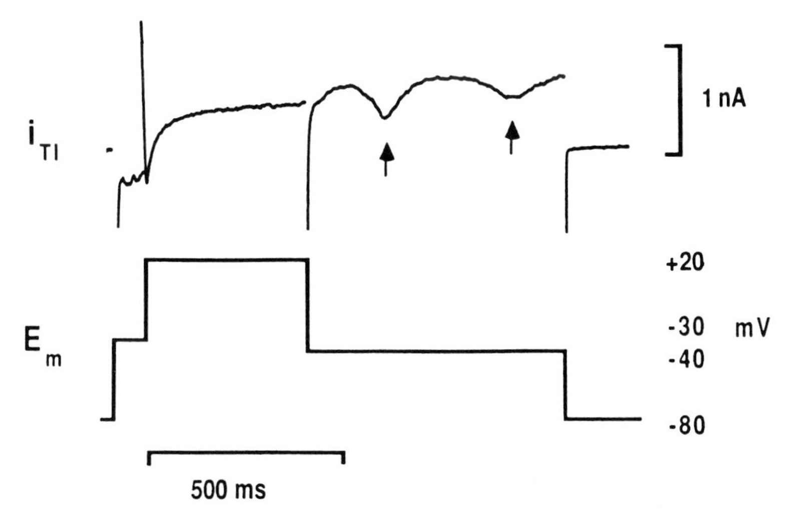

Fig. 4. Transient inward currents induced by oxidant stress in isolated rabbit ventricular myocytes. Cells were exposed to photoactivated rose bengal (50 nmole/l) for 5 min and currents were recorded using the whole-cell voltage clamp technique. The upper trace shows the currents recorded during the voltage protocol shown in the lower trace. The arrows indicate the transient inward currents induced by this protocol.

A similar oscillatory current to those shown in Fig. 4 has been previously described in cells calcium loaded by other means such as Na-pump blockade with ouabain (Kass *et al.*, 1978) or low extracellular potassium (Fedida *et al.*, 1987). The transient inward current induced by oxidant stress was shown to underlie the oscillatory fluctuations in membrane potential and automaticity previously described (Matsuura & Shattock 1989*a*,*b*; 1991). The current–voltage relationship of the rose bengal-induced i_{TI} remained inward even at positive potentials (+40 mV) and the current was almost completely abolished by replacement of extracellular sodium with lithium (Matsuura & Shattock, 1991). This would suggest that, under these conditions, i_{TI} is carried almost entirely by electrogenic Na/Ca exchange ($i_{Na/Ca}$) and not by a calcium-activated non-selective channel (i_{ns}) (Colquhoun *et al.*, 1981; Fedida *et al.*, 1987).

The background conductance

Changes in the time-independent background conductance can have an important influence on membrane potential, electrical stability and susceptibility to arrhythmias (Noma *et al.*, 1984). Fig. 5 shows quasi-steady-state current–voltage (*I*–*V*) relationships obtained under control conditions and during oxidant stress recorded in isolated rabbit ventricular cells.

It is clear that oxidant stress shifts the 'N'-shaped current voltage relationship in an inward direction and alters the slope-conductance at the level of the resting membrane potential. The inward shift in the *I*–*V* relationship causes the relationship to cross the voltage axis at 3 points (−70, −40 and −10 mV) and, at two of these intercepts, the relationship has a positive slope. This would indicate that, in addition to −70 mV, −10 mV may be a stable resting membrane potential. This observation provides an explanation for the oxidant stress-induced prolongation of action potential and the occasional failure to repolarise seen in membrane potential studies. The slope-conductance at the level of the resting membrane potential was decreased from a control value of 40±7 nS to 25±5 nS during oxidant stress indicating a possible reduction of the inward rectifier (i_{k1}). This decrease in slope-conductance would tend to make the cell more vulnerable to arrhythmogenic inward currents and to the induction of automaticity.

Conclusions

It is clear that although many mechanisms contribute to reperfusion-induced cellular injury and arrhythmias, oxidant stress may play a signifi-

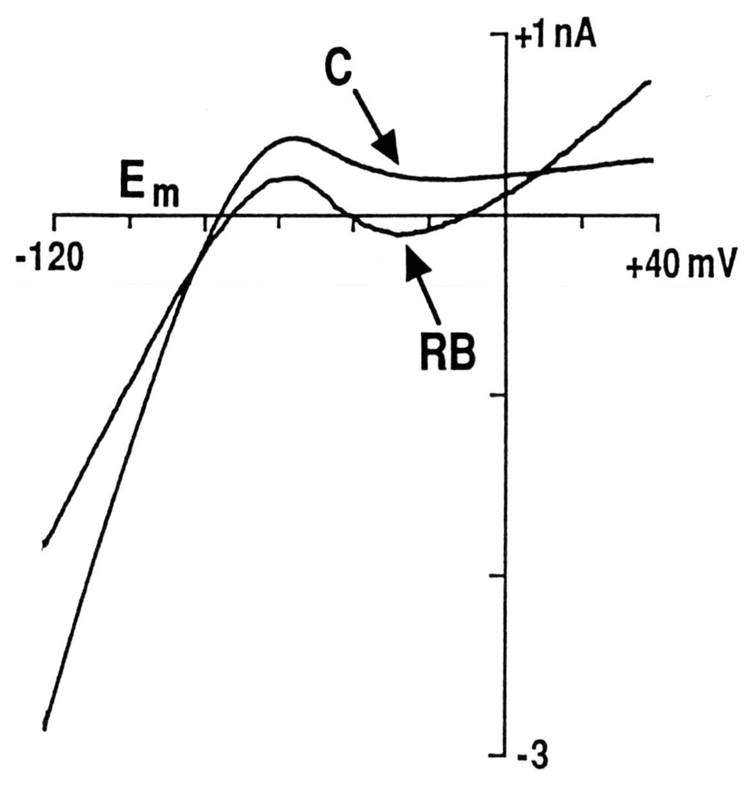

Fig. 5. Changes in background conductance induced by oxidant stress. Currents were recorded under quasi-steady state conditions using a slow ramp-pulse protocol (+50 to −120 mV). Note: oxidant stress shifts the curve inward so that the voltage axis is intercepted at three points (−70, −40 and −10 mV). In addition, the slope conductance at the level of the resting potential (−75 mV) was reduced from 40 nS to 25 nS.

cant role. The sudden burst of free-radical production in early reperfusion is likely to induce a cellular calcium overload and damage to the SR calcium release channel. This cellular calcium overload is not mediated by non-specific membrane damage or by changes in the calcium inward current but appears to be dependent on the transsarcolemmal sodium gradient. This, and the observation that oxidant stress can inhibit the Na-pump current, would suggest that the calcium overload may be due to a rise in intracellular sodium and the activation of the Na/Ca exchange mechanism. As a consequence of the cellular calcium overload and SR damage, oscillatory SR calcium release leads to the activation of an oscillatory transient inward current. In addition to the activation of this

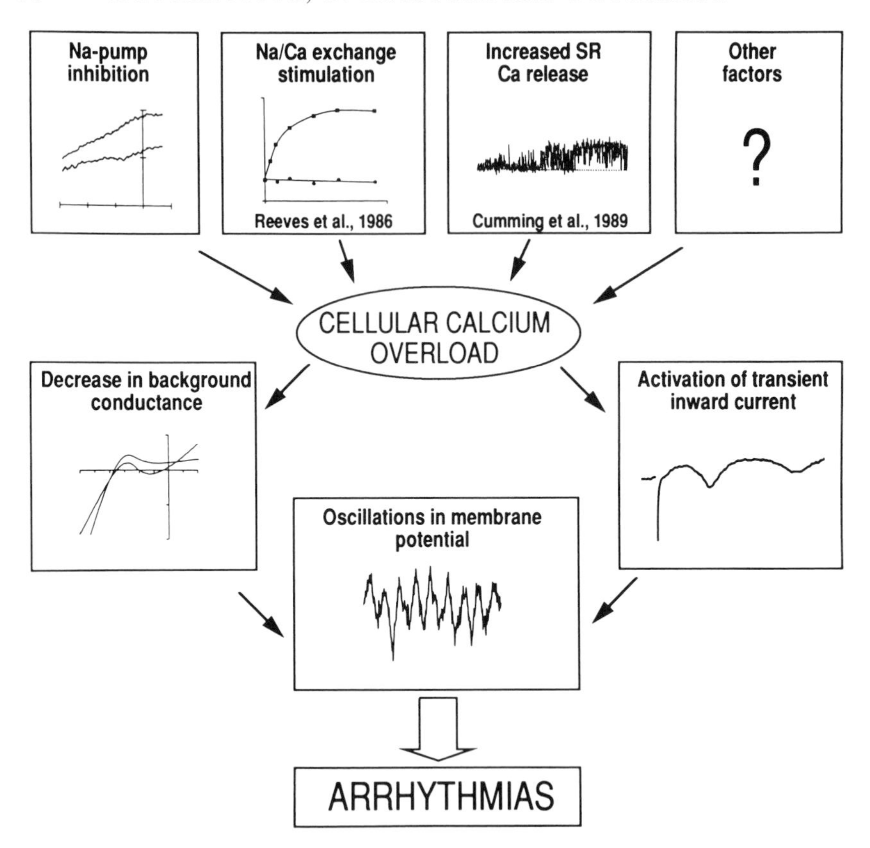

Fig. 6. Diagrammatic representation of the cellular mechanisms underlying oxidant-stress-induced arrhythmias. Na-pump inhibition, stimulation of the Na–Ca exchange mechanism, and increased SR Ca release combine to cause cellular calcium overload. This cellular calcium overload decreases the background conductance and activates an arrhythmogenic transient inward current. The activation of the oscillatory inward current, coupled to the decrease in background conductance, causes oscillations in membrane potential and arrhythmias.

arrhythmogenic transient inward current, the vulnerability of the cell to automaticity and arrhythmias is increased by an oxidant stress-induced decrease in the slope-conductance of the background current. The multiple effects of oxidant stress, and their role in the genesis of arrhythmias, are summarised in Fig. 6.

While it is clear that oxidant stress can cause arrhythmias and may thus provide an important target for therapeutic interventions, it is important to emphasise that many other triggers may exist. The relative contribution of these various arrhythmogenic factors during ischaemia and reperfusion has yet to be fully characterised but it seems likely that oxidant-stress-induced changes in membrane electrophysiology and ionic homeostasis may play an important role.

Acknowledgements

This work was supported by grants from the British Heart Foundation, the National Heart Lung and Blood Institute (HL 37278) and STRUTH.

References

Barrington, P.L., Meier, C.F. & Weglicki, W.B. (1988). Abnormal electrical activity induced by free radical generating systems in isolated cardiocytes. *Journal of Molecular and Cellular Cardiology* **20**, 1163–78.

Bernier, M., Hearse, D.J. & Manning, A.S. (1986). Reperfusion-induced arrhythmias and oxygen-induced free radicals: studies with 'anti-free radical interventions' and a free radical generating system in the isolated perfused rat heart. *Circulation Research* **58**, 331–40.

Bers, D.M. (1985). Calcium influx and sarcoplasmic reticulum Ca release in cardiac muscle activation during postrest recovery. *American Journal of Physiology* **248**, H366–81.

Bindoli, A., Cavallini, L., Rigobello, M.P., Coassin, M. & Di Lisa, F. (1988). Modification of the xanthine converting enzyme of perfused rat heart during ischemia oxidative stress. *Free Radicals in Biology and Medicine* **4**, 163–7.

Bolli, R., Patel, B.S., Jeroudi, M.O., Lai, E.K. & McCay, P.B. (1988). Demonstration of free radical generation in the 'stunned' myocardium in the intact dog with the use of the spin trap alpha-phenyl-N-tert-butyl nitrone. *Federation Proceedings* **2352**, A701.

Borgers, M., Ver Donck, L. & Vandeplassche, G. (1987). Pathophysiology of cardiomyocytes. *Annals of the New York Academy of Science* **522**, 433–53.

Boveris, A. & Chance, B. (1973). The mitochondrial generation of hydrogen peroxide. General properties and effect of hyperbaric oxygen. *Biochemical Journal* **134**, 707–16.

Braunwald, E. & Kloner, R.A. (1982). The stunned myocardium: prolonged post-ischemic ventricular dysfunction. *Circulation* **66**, 1146–9.

Burton, K.P., Morris, A.C., Massey, K.D., Buja, L.M. & Hagler, H.K. (1988). Cellular ionic calcium increases in cultured neonatal rat ventricular myocytes exposed to a free radical generating system. *Journal of Molecular and Cellular Cardiology* **20**, Suppl. V, 59.

Caughey, W.S. & Watkins, J.A. (1985). Oxy radical and peroxide formation by hemoglobin and myoglobin. In *CRC Handbook of Methods for Oxygen Radical Research*, ed. R.A. Greenwald, pp. 95–104. Boca Raton: CRC Press.

Chambers, D.E., Park, D.A., Patterson, G., Roy, R., McCord, J.M., Yoshida, S., Parmley, L.F. & Downey, J.M. (1985). Xanthine oxidase as a source of free radical damage in myocardial ischemia. *Journal of Molecular and Cellular Cardiology* **17**, 145–52.

Coetzee, W.A. & Opie, L.H. (1988). Electrophysiological effects of free oxygen radicals on guinea pig ventricular myocytes. *Journal of Molecular and Cellular Cardiology* **20**, Suppl. V, S 17.

Colquhoun, D., Neher, D., Reuter, H. & Stevens, C.F. (1981). Inward current channels activated by intracellular Ca in cultured cardiac cells. *Nature* **294**, 752–4.

Corr, P.B., Shayman, J.A., Kramer, J.B. & Kipnis, R.J. (1981). Increased alpha-adrenergic receptors in ischemic cat myocardium: a potential mediator of electrophysiological derangements. *Journal of Clinical Investigation* **67**, 1232–326.

Corr, P.B. & Sobel, B.E. (1982). Amphiphilic lipid metabolism and ventricular arrhythmias. In *Early Arrhythmias Resulting From Myocardial Ischaemia*, ed. J.R. Parratt, pp. 199–218. London: MacMillan Press.

Cumming, D., Holmburg, S., Kusama, Y., Shattock, M.J. & Williams, A. (1989). Effect of free radicals on the structure and function of the calcium-release channel from isolated sheep cardiac sarcoplasmic reticulum. *Journal of Physiology* **420**, 88P.

Curtis, M.J. (1989). Regional elevation of extracellular K^+ concentration in the absence of ischemia elicits ventricular arrhythmias: relevance to arrhythmogenesis during ischemia. *Journal of Molecular and Cellular Cardiology* **21**, Suppl. II, 413.

Dennis, S.C., Shattock, M.J., Hearse, D.J., Ball, M.R., Sochor, M. & McLean, P. (1983). Two different metabolic responses to ischemia: inherent variabilty or artifact? *Cardiovascular Research* **17**, 489–98.

Engler, R.L., Dahlgren, M.D., Morris, D., Peterson, M. & Schmid-Schoenbein, G.W. (1986). Role of leucocytes in response to acute myocardial ischemia and reflow in dogs. *American Journal of Physiology* **252**, H314–22.

Fedida, D., Noble, D., Rankin, A.C. & Spindler, A.J. (1987). The arrhythmogenic transient outward current i_{TI} and related contraction in isolated guinea pig ventricular myocytes. *Journal of Physiology* **392**, 523–42.

Garlick, P.B., Davies, M.J., Hearse, D.J. & Slater, T.F. (1987). Direct detection of free radicals in the reperfused rat heart using electron spin resonance spectroscopy. *Circulation Research* **61**, 757–60.

Halliwell, B. & Gutteridge, J.M.C. (1985). *Free Radicals in Biology and Medicine*, pp. 268–76. Oxford: Clarendon Press.

Hammond, B. & Hess, M.L. (1985). The oxygen free radical system:

potential mediator of myocardial injury. *Journal of the American College of Cardiology* **6**, 215–20.

Hearse, D.J. (1989). Free radicals and myocardial injury during ischemia and reperfusion: a short-lived phenomenon? In *Lethal Arrythmias Resulting From Myocardial Ischemia and Infarction*, ed. M.R. Rosen & Y. Palti, pp. 105–15. Boston: Kluwer.

Hearse, D.J. (1991). Reperfusion-induced injury: a possible role for oxidant stress and its manipulation. *Cardiovascular Drugs and Therapy* (in press).

Hearse, D.J., Humphrey, S.M. & Bullock, G.R. (1978). The oxygen paradox and the calcium paradox: two facets of the same problem? *Journal of Molecular and Cellular Cardiology* **10**, 641–68.

Hearse, D.J., Humphrey, S.M. & Chain, E.B. (1973). Abrupt reoxygenation of the anoxic potassium-arrested perfused rat heart: a study of myocardial enzyme release. *Journal of Molecular and Cellular Cardiology* **5**, 395–407.

Hearse, D.J., Humphrey, S.M., Nayler, S.M., Slade, A. & Border, D. (1975). Ultrastructural damage associated with reoxygenation of the anoxis myocardium. Journal of Molecular and Cellular Cardiology **7**, 315–24.

Hearse, D.J., Kusama, Y. & Bernier, M. (1989). Rapid electrophysiological changes leading to arrhythmias in the aerobic rat heart: photosensitization studies with rose bengal-derived reactive oxygen intermediates. *Circulation Research* **65**, 146–53.

Hearse, D.J. & Tosaki, A. (1988). Free radicals and calcium: simultaneous interacting triggers as determinants of vulnerability to reperfusion-induced arrhythmias in the rat heart. *Journal of Molecular and Cellular Cardiology* **20**, 213–23.

Hess, M.L. & Manson, N.H. (1984). Molecular oxygen: friend and foe? *Journal of Molecular and Cellular Cardiology* **16**, 969–85.

Kaneko, M., Beamish, R.E. & Dhalla, N.S. (1989*a*). Depression of heart sarcolemmal Ca^{2+}-pump activity by oxygen free radicals. *American Journal of Physiology* **256**, H368–74.

Kaneko, M., Lee, S.-L., Wolf, C.M. & Dhalla, N.S. (1989*b*). Reduction of calcium channel antagonist binding sites by oxygen free radicals in rat heart. *Journal of Molecular and Cellular Cardiology* **21**, 935–43.

Kass, R.S., Lederer, W.J., Tsien, R.W. & Weingart, R. (1978). Role of calcium ions in transient inward currents and aftercontractions induced by strophanthidin in cardiac Purkinje fibres. *Journal of Physiology* **281**, 187–208.

Kato, M., Kako, K. & Nagano, M. (1988). Effects of oxidation and iron-chelation on sarcolemmal Na^{+}/Ca^{++} exchange. *Journal of Molecular and Cellular Cardiology* **20**, Supp. I, 99.

Kim, M. & Akera, T. (1987). O_2 free radicals: cause of ischemia-reperfusion injury to cardiac Na^{+}-K^{+}-ATPase. *American Journal of Physiology* **252**, H252–7.

Kort, A.A. & Lakatta, E. (1984). Calcium-dependent mechanical oscillations occur spontaneously in unstimulated mammalian cardiac cells. *Circulation Research* **54**, 396–404.

Kramer, J.H., Arroyo, C.M., Dickens, B.F. & Weglicki, W.B. (1987). Spin-trapping evidence that graded myocardial ischemia alters post-ischemic superoxide production. *Free Radicals in Biology and Medicine* **3**, 153–9.

Lee, K.S., Marban, E. & Tsien, R.W. (1985). Inactivation of calcium channels in mammalian heart cells: joint dependence on membrane potential and intracellular calcium. *Journal of Physiology* **364**, 395–411.

Manning, A.S. & Hearse, D.J. (1984). Reperfusion-induced arrhythmias: mechanisms and prevention. *Journal of Molecular and Cellular Cardiology* **16**, 497–518.

Matsuoka, T., Kaminishi, T., Kato, M. & Kako, K.J. (1987). Prevention of free radical-induced dysfunction of membrane-bound (Na-K) ATPase and heart cell damage. *Journal of Molecular and Cellular Cardiology* **19**, 232.

Matsuura, H. & Shattock, M.J. (1989*a*). Functional and electrophysiological effects of reactive oxygen intermediates on isolated rat ventricular muscle. *Journal of Molecular and Cellular Cardiology* **21**, Suppl. II, 386.

Matsuura, H. & Shattock, M.J. (1989*b*). On the mechanism of the arrhythmogenic action of reactive oxygen intermediates in isolated rabbit ventricular cells. *Circulation* **80**, Suppl. II, 118.

Matsuura, H. & Shattock, M.J. (1991). Membrane potential fluctuations and transient inward currents induced by reactive oxygen intermediates in isolated rabbit ventricular cells. *Circulation Research* **68**, 319–29.

Misra, H.P., Weglicki, W.B., Abdulla, R. & McCay, P.B. (1984). Identification of carbon centred free radicals during reperfusion injury in ischemic rat heart. *Circulation* **70**, 11–260.

Noma, A., Nakayame, T., Kurachi, Y. & Irisawa, H. (1984). Resting K conductances in pacemaker and non-pacemaker heart cells of the rabbit. *Japanese Journal of Physiology*, **34**, 245–54.

Opie, L.H. (1989). Reperfusion injury and its pharmacological modification. *Circulation* **80**, 1049–62.

Pallandi, R.T., Perry, M.A. & Campbell, T.J. (1987). Proarrhythmic effects of an oxygen-derived free radical generating system on action potentials recorded from guinea pig ventribular myocardium: A possible cause of reperfusion-induced arrhythmias. *Circulation Research* **61**, 50–4.

Reeves, J.P., Bailey, C.A. & Hale, C.C. (1986). Redox modification of sodium–calcium exchange activity in cardiac sarcolemmal vesicles. *Journal of Biological Chemistry* **11**, 4948–55.

Romson, J., Hook, B., Kunkel, S., Abrams, G., Schork, A. & Lucchesi, B.R. (1983). Reduction in the extent of myocardial ischemic injury by neutrophil depletion in the dog. *Circulation* **67**, 1016–23.

Serruys, P.W., Simoons, M.L., Suryapranata, H., Vermeer, F., Wijns, W., van den Brand, M., Bär, F., Zwaan, C., Krauss, X.H., Remme, W.J., Res, J., Verheugt, F.W.A., van Domburg, R., Lubsen, J. & Hugenholtz, P.G. (1986). Preservation of global and regional left ventricular function after thrombolysis in acute myocardial infarction. *Journal of the American College of Cardiology* **7**, 729–35.

Shattock, M.J. & Matsuura, H. (1990). Oxidant stress inhibits Na/K pump current in isolated rabbit ventricular myocytes. *Journal of Molecular and Cellular Cardiology* **22**, Suppl III, PT 70.

Sheridan, D.J., Penkoske, P.A., Sobel, B.E. & Corr, P.B. (1980). Alpha adrenergic contributions to dysrhythmia during myocardial ischemia and reperfusion in cats. *Journal of Clinical Investigation* **65**, 161–71.

Shi, Z.Q., Davidson, A.J. & Tibbetts, G.F. (1989). Effects of active oxygen generated by DTT/Fe^{2+} on cardiac Na^+/Ca^{2+} exchange and membrane permeability to Ca^{2+}. *Journal of Molecular and Cellular Cardiology* **21**, 1009–16.

Tarr, M. & Valenzeno, D. (1989). Modification of cardiac action potential by photosensitizer-generated reactive oxygen. *Journal of Molecular and Cellular Cardiology* **21**, 539–43.

Vandeplassche, G., Bernier, M., Thone, F., Borgers, M., Kusama, Y. & Hearse, D.J. (1990). Singlet oxygen and myocardial injury: ultrastructural, cytochemical and electrocardiographic consequences of photoactivation of rose bengal. *Journal of Molecular and Cellular Cardiology* **22**, 287–301.

Wang, Y.H., Kakar, S.S., Huang, W.-H. & Askari, A. (1988). Sensitivity of cardiac sodium pump to oxygen free radicals. *Federation of the American Societies for Experimental Biology* **2** (5), A1302.

Yamada, M., Hearse, D.J. & Curtis, M.J. (1990). Readmission of O_2 is unnecessary for the initiation of reperfusion arrhythmias: exeunt oxyradicals? *Journal of Molecular and Cellular Cardiology* **22**, Suppl. III, s 46.

Zweier, J.L., Flaherty, J.T. & Weisfeldt, M.L. (1987). Direct measurement of free radical generation following reperfusion of ischemic myocardium. *Proceedings of the National Academy of Science USA* **84**, 1404–7.

C.J. DUNCAN

Biochemical pathways that lead to the release of cytosolic proteins in the perfused rat heart

Cellular damage in both cardiac and skeletal muscles rapidly follows a rise in the intracellular concentration of calcium ($[Ca^{2+}]_i$) and these damaged cells share a number of features that are summarised in Table 1 (Duncan, 1978). In particular, the myofilament apparatus undergoes characteristic degradative change (Duncan & Jackson, 1987; Rudge & Duncan, 1980) and the sarcolemma is damaged with the consequent efflux of cytosolic proteins, usually conveniently measured as the concentration of creatine kinase (CK) in the surrounding medium or perfusate (Duncan & Jackson, 1987). These two damage pathways can be separately stimulated and act independently, although they have several features in common (Duncan & Jackson, 1987). The other characteristic features of damage in muscle cells listed in Table 1 are probably secondary events that are consequent upon the marked rise in $[Ca^{2+}]_i$ which triggers contraction and hence the consumption of energy stores and a rise in body temperature. Ca^{2+} uptake and excessive ATP hydrolysis result in the *eventual* fall in pH_i, and Ca^{2+} uptake by the mitochondria

Table 1. *Features of rapid muscle damage following a rise in* $[Ca^{2+}]_i$

Ultrastructural damage of myofilament apparatus
Release of cytosolic proteins (LDH, CK)
pH_i falls ($2H^+/Ca^{2+}$ exchange, ATP hydrolysis)
Consumption of high energy stores
Contraction, tension development, hypercontraction (because $[Ca^{2+}]_i$ rises)
Rise in temperature (because of contraction)
Speed of effect
Change in sarcolemma resting potential
Increase in free lysosomal hydrolases
Release of prostaglandins
Release of leukotrienes
Mitochondrial septation

triggers their septation and subdivision (Duncan, 1988*a*). Ca^{2+} also activates phospholipase A_2, with the consequent stimulation of lipoxygenase and cycloxygenase pathways and the release of leukotrienes and prostaglandins.

Of these various damage events, perhaps the most studied has been the pathway that culminates in the release of CK in the isolated, perfused mammalian heart. It has been accepted that Ca^{2+} has a role in the genesis of these events, but there has been increasing interest in a possible major role for active oxygen metabolites which have been detected in the perfusate from a damaged heart (Bolli *et al.*, 1988; Zweier, 1988). In particular, interest has focussed on the possibility of protecting the reperfused ischaemic heart by including superoxide dismutase and catalase at the moment of reperfusion. The results have been conflicting but a clear reduction in the extent of the damage by the use of the oxygen radical scavengers has been reported in many of these studies (Hearse & Tosaki, 1988; Bolli *et al.*, 1989). This paper presents a synthesis of the experimental evidence concerning the breakdown in the integrity of the sarcolemma in cardiac muscle and attempts to assess the respective roles of Ca^{2+} and oxygen radicals in the underlying biochemical pathways.

The Ca^{2+}- and O_2-paradoxes are familiar phenomena in cardiac physiology; perfusion of the isolated rat heart with Ca^{2+}-free saline for 5 min or with anoxic saline for 30 min causes no overt damage, but CK release begins within 1–2 min when Ca^{2+} or O_2 are returned to the perfusion medium (hence the paradox). The sequences of the paradoxes are illustrated in Fig. 1, which emphasizes that each is subdivided into an initial, priming phase that is followed by the damage phase. The features of particular interest in both paradoxes are the rapidity of the onset of CK release in Phase II, the marked sensitivity of Phase I to lowering the temperature to 30 °C and the relative insensitivity to temperature in Phase II (Hearse, Humphrey & Bullock, 1978*b*). Do common biochemical pathways underlie the two paradoxes?

Almost all experiments have been carried out using the conventional Langendorff-perfused heart (with either constant pressure or constant flow) and monitoring CK efflux into the perfusate, but additional information has been obtained by the use of isolated cardiac muscle cells.

The calcium paradox of the heart

A number of suggestions have been advanced for the damage mechanism in Phase I of the Ca^{2+}-paradox, including separation of the intercalated discs so that the cells pull themselves apart on contraction and the separa-

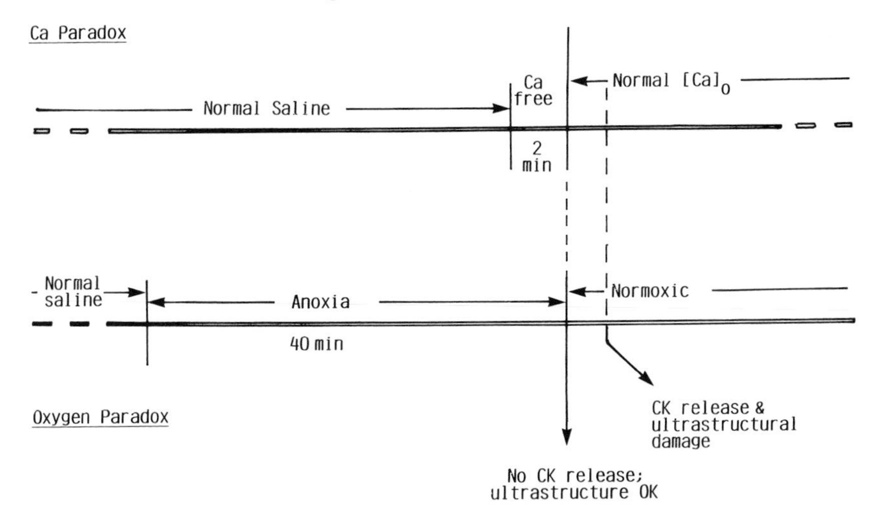

Fig. 1. Comparison of the sequence of events in the protocols for the Ca^{2+}- and O_2-paradoxes. The cells are undamaged until extracellular Ca^{2+} or O_2 are returned, whereupon CK release is detectable within 90 s and the failure of the contractile machinery quickly follows.

tion of the external lamina from the surface of the sarcolemma, resulting in the formation of fluid-filled blebs (see Ruigrok, 1982).

However, long exposures to Ca^{2+}-free solutions containing EGTA are not necessary; perfusion with normal salines lacking Ca^{2+} for 90, 120 or 180 s are sufficient to produce a complete Ca^{2+}-paradox with an efflux of CK beginning within 60 s of reperfusion with normal saline. Since part of the time of Ca^{2+}-free perfusion will be needed for a complete exchange of media within the heart, an exposure of the sarcolemma for 60 s during Phase I is probably sufficient to initiate the damage sequence. Nor does the external medium need to be completely free of Ca^{2+}; the critical concentration, using Ca^{2+}–EGTA buffers, is approximately 8×10^{-7}M when in some hearts a full paradox is obtained, whereas in others no CK is released (Daniels & Duncan, unpublished). Complete protection of the Ca^{2+}-paradox is obtained when phase I is carried out at 30 °C or at lower temperatures (Hearse *et al.*, 1978*b*), suggesting the importance of a phase change in the membrane phospholipids and of the mobility of membrane proteins. It is therefore unlikely that major disruption of the sarcolemma occurs during the brief Ca^{2+}-free period, but rather that some molecular perturbation occurs in the membrane phospholipids and proteins which triggers subsequent events. There is evidence from freeze fracture data that Ca^{2+} depletion causes an apparent decrease in intramembrane particles on the P face within 5 min because of a reorientation in the plane of the bilayer and these changes in the molecular organisation of the sarcolemma are highly temperature-sensitive (Frank *et al.*, 1982).

When extracellular Ca^{2+} is removed there is an accompanying marked rise in intracellular Na^+; Chapman & Tunstall (1987) summarise studies with ^{22}Na or with ion-sensitive microelectrodes using a variety of vertebrate cardiac muscle cells and in all cases the half-maximal Na^+-loading occurs at a $[Ca^{2+}]_o$ of about 1 μM, with significant loading at 10 μM. Some Na^+-loading therefore occurs at levels of $[Ca^{2+}]_o$ that are too great for the Ca^{2+}-paradox to occur. Substances that block this rise in $[Na^+]_i$ reduce or inhibit subsequent CK release.

CK release is clearly initiated in Phase II by the return of extracellular Ca and when the $[Ca^{2+}]$ of the reperfusion medium is reduced from the normal 2.55 mM to 0.1 mM the subsequent damage is dramatically reduced (Daniels & Duncan, unpublished). It is the entry of Ca^{2+} and the rise in $[Ca^{2+}]_i$ that trigger sarcolemma breakdown. Total cellular Ca^{2+} may increase 5- to 40-fold (Goshima, Wakabayashi & Masuda, 1980; Nayler *et al.*, 1984).

The following sequence of events can be suggested for the Ca^{2+}-paradox and the details of some of the stages are amplified below.

Phase I – Removal of Ca_o^{2+}

Stage Ia	Molecular perturbation of the sarcolemma
Stage Ib	Activation of a regulatory protein kinase C
Stage Ic	Activation of a molecular complex (designated 'X'). This stage is temperature-dependent
Stage Id	Activation of a Na^+/H^+ antiporter, H^+ efflux, Na^+ influx

Phase II – Return of Ca_o^{2+}

Stage IIa	Ca^{2+} entry via Na^+/Ca^{2+} exchange in 'reverse mode'
Stage IIb	*Intracellular* Ca^{2+} fully activates X
Stage IIc	Damage to sarcolemma proteins and CK efflux

Thus, in Phase I, both X and the antiporter are activated and Ca^{2+} enters in Phase II by exchange with Na^+ when it is returned extracellularly.

Oxygen radicals and the calcium paradox

If oxygen radicals were implicated in the Ca^{2+}-paradox, the most probable points of action would be the modification of cation movements

(Stages Id and IIa) and the damage to sarcolemma proteins (Stage IIc). Substantial protection has been reported when the heart is perfused with superoxide dismutase plus catalase before and during the Ca^{2+}-paradox (Ashraf, 1987). However, no reduction in CK release was found when the following range of oxygen-radical scavengers was included in the perfusion medium throughout the procedure of the Ca^{2+}-paradox, including a 20 min preincubation period: dimethylthiourea, mannitol, desferrioxamine, or superoxide dismutase plus catalase (Daniels & Duncan, 1989). The Ca^{2+}-paradox in rabbit hearts did not produce a decrease in the tissue content of protein sulphydryl groups, suggesting that no oxidative damage had occurred (Ferrari *et al.*, 1989) and it is concluded that oxygen radicals do not have a major role in the genesis of damage.

Equally persuasive is the evidence obtained when the Ca^{2+}-paradox is carried out under anoxia. The saline was heavily gassed with 95% N_2 + 5% CO_2 and then perfused for 20 min before removal of extracellular Ca^{2+}; no protection was found and, on the contrary, CK release was slightly exacerbated (Daniels & Duncan, unpublished) and the results confirm that oxygen radicals are not implicated. Furthermore, CK release begins during Phase I under anoxia (i.e. before the return of extracellular Ca^{2+}) and, when glucose is also omitted, CK release is greatly increased. Release of CK begins within 1 min of the priming action of the removal of extracellular Ca^{2+} under anoxia. Thus, if the breakdown in the integrity of the sarcolemma is caused by the localised release of active oxygen radicals, *in vivo*, the system can readily switch to the direct transferance of electrons to key membrane proteins.

The rise in $[Ca^{2+}]_i$

Such findings suggest that anoxic conditions, particularly in the absence of extracellular glucose, cause the rapid depletion of high-energy stores in this metabolically active tissue, the inhibition of Ca^{2+}-transport mechanisms and the consequent progressive rise in $[Ca^{2+}]_i$. This rise in $[Ca^{2+}]_i$ therefore substitutes for the massive Ca^{2+} entry in Stage IIb of the Ca^{2+}-paradox, but no CK release occurs until X is activated by removal of extracellular Ca^{2+} (Stages Ia to Ic).

Rabbit and guinea pig hearts differ from the rat heart preparation in that they release substantial amounts of CK from the start of the Ca^{2+}-free perfusion (Hearse *et al.*, 1978*a*) and presumably intracellular conditions in these preparations resemble the anoxic rat heart; $[Ca^{2+}]_i$ is high and CK begins when the cells are activated by removal of extracellular Ca^{2+}.

$[Ca^{2+}]_i$ can also be raised experimentally by perfusing the rat heart with caffeine or 2,4-dinitrophenol and both treatments give the similar results (Daniels & Duncan, unpublished). Caffeine causes the release of Ca^{2+} from the sarcoplasmic reticulum and dinitrophenol uncouples mitochondria, thereby releasing accumulated Ca^{2+}. Inclusion of either of these agents in saline containing the normal concentration of Ca^{2+} did not cause the release of CK after prolonged perfusion. However, when the heart is primed by an initial perfusion with Ca^{2+}-free saline to which is then added caffeine or dinitrophenol, CK release begins within 2 min of the addition and the total efflux is comparable to that recorded with a standard Ca^{2+}-paradox. Perfusion at 28 °C provides complete protection (Ganote & Sims, 1984; Daniels & Duncan, unpublished).

Since the experiments are conducted with Ca^{2+}-free media, caffeine is clearly not acting to promote Ca^{2+}-influx but is promoting release of Ca^{2+} from intracellular sites and this rise in $[Ca^{2+}]_i$ is substituting for Stage IIb of the paradox, confirming the view that it is the intracellular concentration of Ca^{2+} and not the flux of Ca^{2+} across the sarcolemma that finally precipitates CK release. However, again, the system needs activation via the removal of extracellular Ca^{2+} and the absolute sensitivity of activation to temperatures below 30 °C suggests that this process is strictly comparable to Stage Ic of the Ca^{2+}-paradox. It is evident that brief removal of extracellular Ca^{2+} is a potent activator of the damage system.

Action of amiloride

Amiloride is a specific inhibitor of the Na^+/H^+ antiporter of the plasma membrane (Papa, Lorusso & Capuano, 1988). The divalent cation ionophore A23187 administered together with 12-*0*-tetradecanoyl-phorbol-13-acetate (TPA) caused a rise in pH_i in isolated mouse cardiac cells by activating amiloride-sensitive Na^+/H^+ exchange and also produced a rise in $[Ca^{2+}]_i$ above that seen with A23187 alone. TPA also enhanced CK release from hypoxic myocardiac cells and all these effects were suppressed by the addition of 1 mM amiloride (Ikeda *et al.*, 1988). Amiloride also completely inhibited CK release in the Ca^{2+}-paradox and with caffeine in the perfused rat heart (Daniels & Duncan, unpublished) and such findings suggest that the priming of X in Phase I is associated with the activation of a Na^+/H^+ antiporter, thereby producing an efflux of H^+ and the observed influx of Na^+ in Stage Id (Chapman & Tunstall, 1987). The subsequent Ca^{2+}/Na^+ exchange operating in 'reverse mode' causes Ca^{2+} entry in Stage IIa (Fig. 2).

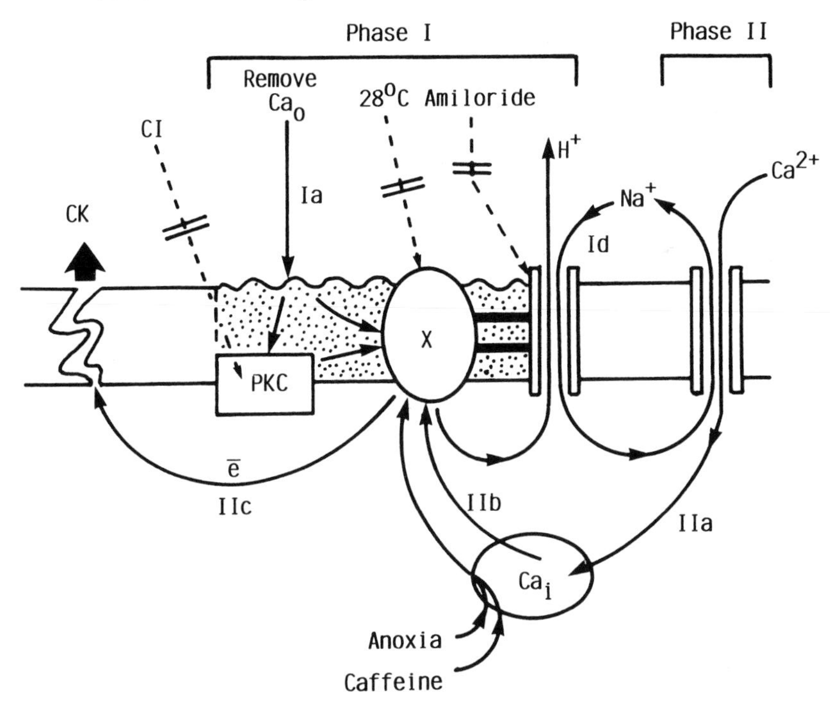

Fig. 2. Events during the Ca^{2+}-paradox. Removal of extracellular Ca^{2+} (Phase I) causes the activation of a transmembrane oxido-reductase complex (X) and the generation of H^+ which are exchanged for extracellular Na^+ via the antiporter. The return of extracellular Ca^{2+} permits Na^+/Ca^{2+} exchange, a rise in $[Ca^{2+}]_i$ and the further activation of X by positive feedback. The Stages (Ia to IIc) described in the text are identified on the figure.

The action of Ca^{2+} in Stage IIc of the Ca^{2+}-paradox

How does the rise in $[Ca^{2+}]_i$ in Stage IIb trigger CK release in Stage IIc? A number of mechanisms could be suggested. Firstly, Ca^{2+}-activation of a phospholipase A_2 (PLA_2) would generate arachidonic acid which acts as the substrate for the lipoxygenase and cycloxygenase pathways, generating leukotrienes and prostaglandins respectively and both these classes of active compounds have been detected in damaged cardiac muscles (Karmazyn & Moffat, 1984; Karmazyn, 1987; Trevethick *et al.*, 1989). Furthermore, mepacrine (a PLA_2 inhibitor) and nordihydroguaretic acid (a lipoxygenase inhibitor) markedly inhibit CK release initiated by a rise in $[Ca^{2+}]_i$ in isolated skeletal muscle (Duncan & Jackson,

1987). However, neither mepacrine nor nordihydroguaretic acid protect against the Ca^{2+}-paradox (Duncan, 1988*b*; Daniels & Duncan, 1989) and we conclude that although PLA_2 is activated under extreme damage with high levels of $[Ca^{2+}]_i$ its metabolites do not have a major role in the events of the Ca^{2+}-paradox.

Secondly, Ca^{2+} could activate calpain proteases (Duncan, Smith & Greenaway, 1979), or even bring about the release of lysosomal acid hydrolases (Duncan, Greenaway & Smith, 1980; Duncan & Rudge, 1988; Duncan, 1989). However, $[Ca^{2+}]_i$ rises to stimulating levels during caffeine, dinitrophenol or anoxic perfusion so that proteases would be activated but no CK is released until extracellular Ca^{2+} is removed.

It is concluded that the action of the rise in intracellular Ca^{2+} produced by caffeine or during the oxygen paradox (see below) acts synergistically with the activation of X in Phase I and it is suggested that the rise in intracellular Ca^{2+} operates as a positive feedback, further activating X during Phase II of the Ca^{2+}-paradox.

The oxygen paradox

CK release is rapidly switched on in the oxygen paradox by the return of O_2 to the anoxic heart. Active oxygen metabolites have been detected following reperfusion of the ischaemic heart (Zweier, 1988; Bolli *et al.*, 1988; 1989; Baker & Kalyanaraman, 1989), protection has been claimed by perfusion with oxygen radical scavengers (Vander Heide, Sobotka & Ganote, 1987; Hearse & Tosaki, 1988) and oxygen radicals appear to be prime candidates for a direct role in producing sarcolemma damage. The oxygen paradox is carried out with normal $[Ca^{2+}]_o$ throughout so that, unlike the Ca^{2+}-paradox, there is no priming via molecular perturbation and presumably Na^+ entry is not involved. Furthermore, if O_2 is returned in saline with only 0.1 mM $[Ca^{2+}]_o$ no reduction in CK release was found (Daniels & Duncan, unpublished) so that, again unlike the Ca^{2+}-paradox, Ca^{2+} entry is not involved. Prolonged anoxic perfusion will, as in the experiments discussed above, cause a rise in $[Ca^{2+}]_i$ and the sequence of events in the oxygen paradox may be summarised as follows.

Phase I – Anoxic perfusion, minus extracellular glucose

Stage Ia Loss of high-energy phosphates, build-up of $[Ca^{2+}]_i$

Stage Ib Priming of X by elevated $[Ca^{2+}]_i$. This stage is critically temperature-sensitive

Phase II – Return of O_2

Stage IIa O_2 and raised $[Ca^{2+}]_i$ activate X synergistically

Stage IIb Damage to sarcolemma proteins and CK release

Table 2. *Characteristics of caffeine-induced damage and the Ca^{2+}- and O_2-paradoxes*

	Caffeine	Ca^{2+}-paradox	O_2-paradox
1. Obligatorily dependent on a rise in $[Ca^{2+}]_i$	Yes	Yes (Phase II)	Yes (Phase I)
2. Translation temp at 30 °C	Yes	Yes (Phase I)	Yes (Phase I)
3. Inhibition by amiloride	Yes	Yes	No
4. CI inhibition	Yes	Yes	No
5. Dependent on activation by removal of extracellular Ca^{2+}	Yes	Yes	No
6. Dependent on Ca^{2+}	No	Yes	No

It is evident that the underlying events during the two phases of the paradoxes are different and these differences provide clues concerning the details of the biochemical mechanisms. $[Ca^{2+}]_i$ rises and activates X in Phase II of the Ca^{2+}-paradox but in Phase I of the oxygen paradox. Since molecular perturbation of the sarcolemma is not involved in the oxygen paradox, the suggested sequence of events is much simpler, and yet Phase I of the oxygen paradox shows an identical critical temperature-sensitivity to that of the Ca^{2+}-paradox (Hearse *et al.*, 1978*b*), indicating the same dependence on the transition temperature of the sarcolemma and demonstrating that it is not the molecular perturbation produced by removal of extracellular Ca^{2+}, nor the possible activation of a regulatory protein kinase C (see below) in the Ca^{2+}-paradox that is temperature-sensitive. It is therefore the activation and operation of X itself that is dependent on mobility within the membrane bilayer.

Amiloride had no inhibitory effect on the release of CK in the oxygen paradox (Daniels & Duncan, unpublished), in contrast with the Ca^{2+}-paradox where the Na^+/H^+ antiporter is believed to be driven by X and to be responsible for Na^+ entry in Stage Id, Na^+ subsequently being removed from the cell in exchange for Ca^{2+} entry. This finding is in agreement with no important role of Ca^{2+} influx in the oxygen paradox, where $[Ca^{2+}]_i$ is already raised during anoxia in Phase I, and demonstrates that X and the Na^+/H^+ antiporter exchange are operationally separate.

The details of caffeine-induced damage and the O_2- and Ca^{2+}-paradoxes are summarised and compared in Table 2.

Action of O_2 in the oxygen paradox

Perfusion of rat hearts during the oxygen paradox with dimethylthiourea, allopurinol or catalase did not protect against CK release (Vander Heide *et al.*, 1987). In contrast, CK release upon re-oxygenation of the isolated rabbit heart was almost completely inhibited by the inclusion of either catalase, allopurinol or desferrioxamine to the medium although this was not accompanied by a parallel preservation of ventricular function. Superoxide dismutase alone did not reduce CK release, and it was concluded that the production of H_2O_2, which then formed hydroxyl radicals, was an important component of sarcolemma damage (Myers *et al.*, 1985).

These latter experiments suggest that the return of molecular oxygen to the anoxic hearts provides an electron acceptor for the production of superoxide radicals from which more active oxygen radicals are generated which cause sarcolemma breakdown. However, our studies in which dimethylthiourea, desferrioxamine or catalase were included in the oxygen paradox of the rat heart, provided no evidence of inhibition of CK release (Daniels & Duncan, unpublished), corresponding with the results of Vander Heide *et al.* (1987) and to the absence of protection in the Ca^{2+}-paradox.

It is concluded that the return of O_2 does not trigger the production of superoxides and other active oxygen metabolites in the highly artificial conditions of the oxygen paradox of the perfused rat heart, and it is suggested that it acts by the stimulation of a transmembrane oxidase, perhaps directly or perhaps by the reactivation of the pentose phosphate pathway that drives a sarcolemma NAD(P)H dehydrogenase (see below and Fig. 3).

Does protein kinase C have a regulatory role?

As described above, the protein kinase C activator TPA stimulated the Na^+/H^+ antiporter, H^+ efflux, a rise in $[Ca^{2+}]_i$ and CK release in myocardial cells, but only when $[Ca^{2+}]_i$ was initially raised by A23187 or anoxia (Ikeda *et al.*, 1988), suggesting that protein kinase C may have a role in regulating the Ca^{2+}-paradox. These conditions were simulated in the perfused rat heart by an initial period of anoxia followed by administration of TPA (in lieu of the removal of extracellular Ca^{2+}) and a small release of CK was recorded. Perfusion of the protein kinase C inhibitor 1-(5-isoquinolinylsulphonyl)-2-methyl-piperazine (CI) during the standard Ca^{2+}-paradox provided complete protection. Although CI is not a specific inhibitor, the results provide some confirmatory evidence

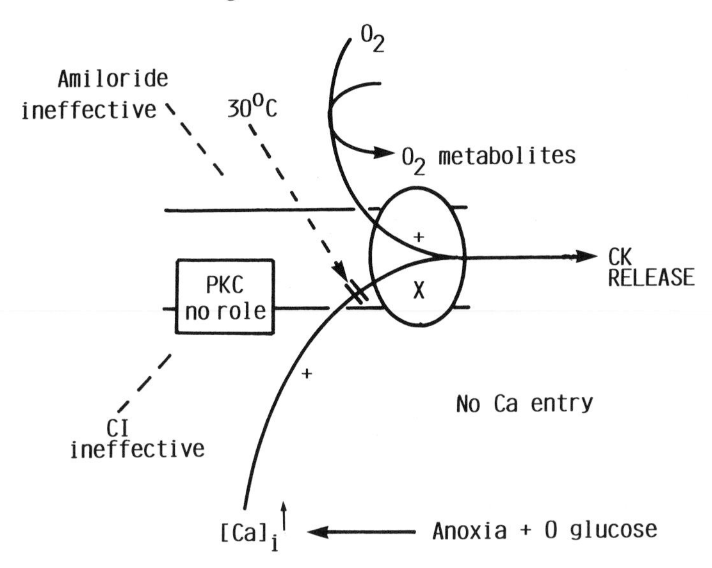

Fig. 3. Events during the O_2-paradox. Anoxia indirectly causes the rise in $[Ca^{2+}]_i$ that activates the transmembrane oxido-reductase complex (X) during Phase I; the damage system and CK release are switched on when O_2 is returned (Phase II). PKC apparently has no regulatory role and amiloride does not prevent CK release.

that regulation may be achieved by phosphorylation of the molecular complex of X and the Na^+/H^+ antiporter via protein kinase C. However, CI provides no protection against the oxygen paradox, showing that the synergistic activation of the damage system by O_2 and $[Ca^{2+}]_i$ is not dependent on protein kinase C activation (Daniels & Duncan, unpublished).

The fundamental differences between the two paradoxes are summarised in Table 2 and a suggested outline for the sequences of events are given in Figs 2 and 3. Activation of X is obligatorily dependent on a rise in $[Ca^{2+}]_i$ in conjunction with an activating event at the sarcolemma although, under prolonged anoxia, a rise in $[Ca^{2+}]_i$ alone may be adequate.

Properties of the molecular complex (X) that causes the damage to the sarcolemma proteins

From the foregoing and Figs 2 and 3 it is possible to assemble a list of features and properties that characterise X:

(i) directly stimulated by Ca^{2+} (as in the oxygen paradox)
(ii) linked to the Na^+/H^+ antiporter with H^+ efflux and Na^+ influx following removal of extracellular Ca^{2+}
(iii) stimulated directly or indirectly by O_2
(iv) dependent on mobility within the membrane bilayer
(v) capable, when activated, of producing membrane damage
(vi) capable of operating under N_2
(vii) probably capable of generating oxygen radicals.

The suggestion has been advanced (Duncan, 1990) that a suitable candidate would be a vectorially-organised, transmembrane NAD(P)H oxido-reductase generating H^+ and electrons which has features in common with a comparable molecular complex present in neutrophils where O_2 acts as the electron acceptor, thereby generating $O_2^{\overline{\cdot}}$ (see Edwards, this volume). Such enzyme complexes are present in the plasma membranes of all cells and have been found in the heart (Crane *et al.*, 1985), but it is difficult to demonstrate conclusively their central role in effecting sarcolemma breakdown and CK release because of the lack of specific inhibitors.

The NADPH oxidase complex of neutrophils has a number of features that correspond with the events during the genesis of CK release in the heart:

(i) sensitivity to Ca^{2+}
(ii) sensitivity to membrane perturbation, for example, with detergents or by changes in the tonicity of the medium (hyperosmolar solutions modify CK release from rat hearts; Ganote, Iannotti & Kaltenbach, 1978)
(iii) transmembrane localisation
(iv) activated by O_2
(v) ability to generate superoxides and O_2 metabolites
(vi) complex stimulation by TPA; $O_2^{\overline{\cdot}}$ production has TPA-sensitive and -insensitive fractions
(vii) temperature-sensitivity, especially activation by arachidonic acid or inhibition of protein kinase C stimulation
(viii) Na^+/H^+ antiporter activity and H^+ efflux linked to $O_2^{\overline{\cdot}}$ production
(ix) the NADPH oxidase of neutrophils is electrogenic and associated with a H^+ channel (Henderson, Chappell & Jones, 1987; Takanaka & O'Brien, 1988)
(x) Na^+/Ca^{2+} exchange is now known to be implicated in the

full activation of superoxide production by neutrophils (Simchowitz, Foy & Cragoe, 1990)

(xi) halothane activates protein kinase C and $O_2^{\cdot-}$ production in neutrophils by 500%; Tsuchiya *et al.* (1988) (compare with the action of halothane on muscle, Arthur & Duthie, this volume, and hepatocytes, Nicotera *et al.*, this volume).

In summary, it is suggested that when stimulated by brief removal of extracellular Ca^{2+} the NAD(P)H oxido-reductase generates H^+ and electrons; it is closely linked both spatially and functionally to a Na^+/H^+ antiporter and, when extracellular Ca^{2+} is returned, Na^+/Ca^{2+} exchange effects the rise in $[Ca^{2+}]_i$ that fully activates the NAD(P)H oxido-reductase generating a flow of electrons. Sulphydryl groups on membrane integral proteins act as electron acceptors, their oxidation causing microlesions in the sarcolemma and CK release. In the oxygen paradox, O_2 and a raised $[Ca^{2+}]_i$ fully activate the NAD(P)H oxidase which generates electrons and directly causes sulphydryl oxidation. Ultrastructural changes in the sarcolemma, with aggregation of intramembrane particles and the formation of vesicles have been described during Phase II of the Ca^{2+}-paradox (Post *et al.*, 1985).

Artificial generation of oxygen radicals

Perfusion of hearts with free-radical-generating systems caused ultrastructural damage and ventricular fibrillation (Bernier, Hearse & Manning, 1986; Ytrehus *et al.*, 1987; Miki *et al.*, 1988) and the anthracycline quinones have a cardiac toxicity associated with the formation of free radicals by one-electron reduction (Davies & Doroshow, 1986). Menadione, another quinone, also causes severe cellular damage and CK release in skeletal muscle (McCall & Duncan, 1990) and its action on the perfused rat heart is currently being studied (Daniels & Duncan, unpublished). Menadione does not cause the release of CK in the heart unless the system is primed by the removal of extracellular Ca^{2+} and these preliminary results show clear parallels with the action of caffeine on the heart. No CK was released when the experiments were repeated under N_2, suggesting that menadione acts by the generation of active oxygen metabolites; however, such findings do not necessarily indicate that free radicals directly cause the oxidation of sulphydryl groups in the sarcolemma that is responsible for CK release. Menadione is known to disrupt the intracellular Ca^{2+} homeostasis and also to activate cytosolic protein kinase C to a high-V_{max} form in hepatocytes by a reduction-sensitive modification of its thiol–disulphide status (Kass, Duddy &

Orrenius, 1989) and it is probable that the primary action of menadione in the heart is to effect the rise in $[Ca^{2+}]_i$ that is apparently obligatory for activation (Table 2); CK release occurs when this is combined with the molecular perturbation associated with the removal of extracellular Ca^{2+}.

In conclusion, oxygen radicals and active oxygen metabolites are generated during cellular damage in cardiac muscle, perhaps as a consequence of the activation of a sarcolemma NAD(P)H oxido-reductase and, when artificially generated, they can cause characteristic ultrastructural damage, although this is almost certainly via an elevation of $[Ca^{2+}]_i$. Nevertheless, there seems to be no evidence to implicate them in a major role in the biochemical pathway underlying CK release, which continues under N_2.

Acknowledgements

I thank Miss Stephanie Daniels who carried out many of the experiments on which this work is based and Miss Susan Scott for assistance in the preparation of the manuscript.

References

Ashraf, M. (1987). Oxygen derived radicals related injury in the heart during calcium paradox. *Virchows Arch. B* **54**, 27–37.

Baker, J.E. & Kalyanaraman, B. (1989). Ischemia-induced changes in myocardial paramagnetic metabolites: implications for intracellular oxy-radical generation. *FEBS Letters* **244**, 311–14.

Bernier, M., Hearse, D.J. & Manning, A.S. (1986). Reperfusion-induced arrhythmias and oxygen-derived free radicals. Studies with 'anti-free radical' interventions and a free radical-generating system in the isolated perfused rat heart. *Circulation Research* **58**, 331–40.

Bolli, R., Patel, B.S., Jeroudi, M.O., Lai, E.K. & McCay, P.B. (1988). Demonstration of free radical generation in 'stunned' myocardium of intact dogs with the use of the spin trap α-phenyl N-tert-butyl nitrone. *Journal of Clinical Investigation* **82**, 476–85.

Bolli, R., Jeroudi, M.O., Patel, B.S., DuBose, C.M., Lai, E.K., Roberts, R. & McCay, P.B. (1989). Direct evidence that oxygen-derived free radicals contribute to postischemic myocardial dysfunction in the intact dog. *Proceedings of the National Academy of Science USA* **86**, 4695–9.

Chapman, R.A. & Tunstall, J. (1987). The calcium paradox of the heart. *Progress in Biophysics and Molecular Biology* **50**, 67–96.

Crane, F.L., Sun, I.L., Clark, M.G., Grebing, C. & Low, H. (1985). Transplasma-membrane redox systems in growth and development. *Biochimica et Biophysica Acta* **811**, 233–64.

Daniels, S. & Duncan, C.J. (1989). Are oxygen-radicals implicated in the calcium paradox? *Biochemical Society Transactions* **17**, 700–7.

Davies, K.J.A. & Doroshow, J.H. (1986). Redox cycling of anthracyclines by cardiac mitochondria. *Journal of Biological Chemistry* **261**, 3060–7.

Duncan, C.J. (1978). Role of intracellular calcium in promoting muscle damage: a strategy for controlling the dystrophic condition. *Experientia* **34**, 1531–5.

Duncan, C.J. (1988*a*). Mitochondrial division in animal cells. In *The Division and Segregation of Organelles*, ed. S.A. Boffey & D. Lloyd, pp. 95–113. Cambridge University Press.

Duncan, C.J. (1988*b*). The role of phospholipase A_2 in calcium-induced damage in cardiac and skeletal muscle. *Cell and Tissue Research* **253**, 457–62.

Duncan, C.J. (1989). The mechanisms that produce rapid and specific damage to the myofilaments of amphibian skeletal muscle. *Muscle and Nerve* **12**, 210–18.

Duncan, C.J. (1990). Biochemical events associated with rapid cellular damage during the oxygen- and calcium-paradoxes of the mammalian heart. *Experientia* **46**, 41–8.

Duncan, C.J., Greenaway, H.C. & Smith, J.L. (1980). 2,4-dinitrophenol, lysosomal breakdown and rapid myofilament degradation in vertebrate skeletal muscle. *Naunyn-Schmiedeberg's Arch. Pharmacol.* **315**, 77–82.

Duncan, C.J. & Jackson, M.J. (1987). Different mechanisms mediate structural changes and intracellular enzyme efflux following damage to skeletal muscle. *Journal of Cell Science* **87**, 183–8.

Duncan, C.J. & Rudge, M.F. (1988). Are lysosomal enzymes involved in rapid damage in vertebrate muscle cells? A study of the separate pathways leading to cellular damage. *Cell and Tissue Research* **253**, 447–55.

Duncan, C.J., Smith, J.L. & Greenaway, H.C. (1979). Failure to protect frog skeletal muscle from ionophore-induced damage by the use of the protease inhibitor leupeptin. *Comparative Biochemistry and Physiology* **63**C, 205–7.

Ferrari, R., Ceconi, C., Curello, S., Cargnoni, A. & Ruigrok, T.J.C. (1989). No evidence of oxygen free radicals-mediated damage during the calcium paradox. *Basic Research in Cardiology* **84**, 396–403.

Frank, J.S., Rich, T.L., Beydler, S. & Kreman, M. (1982). Calcium depletion in rabbit myocardium. Ultrastructure of the sarcolemma and correlation with the calcium paradox. *Circulation Research* **51**, 117–30.

Ganote, C.E., Iannotti, J.P. & Kaltenbach, J.P. (1978). Effects of hyperosmolar solutions of polyethylene glycol, dextran or mannitol on enzyme release from perfused rat hearts. *Journal of Molecular and Cellular Cardiology* **10**, 725–37.

Ganote, C.E. & Sims, M.A. (1984). Parallel temperature dependence of contracture-associated enzyme release due to anoxia, 2,4-dinitrophenol (DNP), or caffeine and the calcium paradox. *American Journal of Pathology* **116**, 94–106.

Goshima, K., Wakabayashi, S. & Masuda, A. (1980). Ionic mechanism of morphological changes of cultured myocardial cells on successive incubation in media without and with Ca^{2+}. *Journal of Molecular and Cellular Cardiology* **12**, 1135–57.

Hearse, D.J., Humphrey, S.M., Boink, A.B.T.J. & Ruigrok, T.J.C. (1978*a*). The calcium paradox: metabolic, electrophysiological, contractile and ultrastructural characteristics in four species. *European Journal of Cardiology* **7**, 241–56.

Hearse, D.J., Humphrey, S.M. & Bullock, G.R. (1978*b*). The oxygen paradox and the calcium paradox: two facets of the same problem. *Journal of Molecular and Cellular Cardiology* **10**, 641–68.

Hearse, D.J. & Tosaki, A. (1988). Free radicals and calcium: simultaneous interacting triggers as determinants of vulnerability to reperfusion-induced arrhythmias in the rat heart. *Journal of Molecular and Cellular Cardiology* **20**, 213–23.

Henderson, L.M., Chappell, J.B. & Jones, O.T.G. (1987). The superoxide-generating NADPH oxidase of human neutrophils is electronic and associated with an H^+ channel. *Biochemical Journal* **246**, 325–9.

Ikeda, U., Arisaka, H., Takayasu, T, Takeda, K., Natsume, T. & Hosoda, S. (1988). Protein kinase C activation aggravates hypoxic myocardial injury by stimulating Na^+/H^+ exchange. *Journal of Molecular and Cellular Cardiology* **20**, 493–500.

Karmazyn, M. (1987). Calcium paradox-evoked release of prostacyclin and immunoreactive leukotriene C4 from rat and guinea-pig hearts. Evidence that endogenous prostaglandins inhibit leukotriene biosynthesis. *Journal of Molecular and Cellular Cardiology* **19**, 221–30.

Karmazyn, M. & Moffat, M.P. (1984). Calcium-ionophore stimulated release of leukotriene C_4-like immunoreactive material from cardiac tissue. *Journal of Molecular and Cellular Cardiology* **16**, 1071–3.

Kass, G.E.N., Duddy, S.K. & Orrenius, S. (1989). Activation of hepatocyte protein kinase C by redox-cycling quinones. *Biochemical Journal* **260**, 499–507.

McCall, K. & Duncan, C.J. (1990). Priming the damage system in mammalian skeletal muscle. *Biochemical Society Transactions* **18**, 608–9.

Miki, S., Ashraf, M., Salka, S. & Sperelaki, N. (1988). Myocardial dysfunction and ultrastructural alterations mediated by oxygen metabolites. *Journal of Molecular and Celluar Cardiology* **20**, 1009–24.

Myers, C.L., Weiss, S.J., Kirsh, M.M. & Shlafer, M. (1985). Involvement of hydrogen peroxide and hydroxyl radical in the 'oxygen paradox': reduction of creatine kinase release by catalase, allopurinol or deferoxamine, but not by superoxide dismutase. *Journal of Molecular and Cellular Cardiology* **17**, 675–84.

Nayler, W.G., Perry, S.E., Elz, J.S. & Daly, M.J. (1984). Calcium, sodium and the calcium paradox. *Circulation Research* **55**, 227–37.

Papa, S., Lorusso, M. & Capuano, F. (1988). pH homeostasis and cell functions and diseases. In *Cell Function and Disease*, ed. L.E. Canedo, L.E. Todd, L. Packer & J. Jaz, pp. 147–57. New York: Plenum.

Post, J.A., Nievelstein, P.F.E.M., Leunissen-Bijvelt, J., Verkleij, A.J. & Ruigrok, T.J.C. (1985). Sarcolemmal disruption during the calcium paradox. *Journal of Molecular and Cellular Cardiology* **17**, 265–73.

Rudge, M.F. & Duncan, C.J. (1980). The experimental induction of ultrastructural damage in cardiac muscle. *Experientia* **36**, 992–3.

Ruigrok, T.J.C. (1982). The calcium paradox: mechanisms and clinical relevance. In *The Role of Calcium in Biological Systems*, vol. III, ed. L.J. Anghileri & A.M. Tuffet-Anghileri, pp. 133–41, Bocca Rakon: CRC.

Simchowitz, L., Foy, M.A. & Cragoe Jr., E.J. (1990). A role for Na^+/Ca^{2+} exchange in the generation of superoxide radicals by human neutrophils. *Journal of Biological Chemistry* **265**, 13449–56.

Takanaka, K. & O'Brien, P.J. (1988). Proton release associated with respiratory burst of polymorphonuclear leukocytes. *Journal of Biochemistry, Tokyo* **103**, 656–60.

Trevethick, M.A., Brown, A.K., Wright, G. & Strong, P. (1989). Cyclo-oxygenase inhibition does not unmask leukotriene release during ischaemia–reperfusion of the rat heart *in vitro*. *Biochemical Pharmacology* **38**, 377–9.

Tsuchiya, M., Okimasu, E., Ueda, W., Hirakawa, M. & Utsumi, K. (1988). Halothane, an inhalation anesthetic, activates protein kinase C and superoxide generation by neutrophils. *FEBS Letters* **242**, 101–5.

Vander Heide, R.S., Sobotka, P.A. & Ganote, C.E. (1987). Effects of the free radical scavenger DMTU and mannitol on the oxygen paradox in perfused rat hearts. *Journal of Molecular and Cellular Cardiology* **19**, 615–25.

Ytrehus, K., Myklebust, R., Olsen, R. & Mjos, O.D. (1987). Ultrastructural changes induced in the isolated rat heart by enzymatically generated oxygen radicals. *Journal of Molecular and Cellular Cardiology* **19**, 379–89.

Zweier, J.L. (1988). Measurement of superoxide-derived free radicals in the reperfused heart. Evidence for a free radical mechanism of reperfusion injury. *Journal of Biological Chemistry* **263**, 1353–7.

JOHN R. ARTHUR and GARRY G. DUTHIE

Malignant hyperthermia: the roles of free radicals and calcium?

Introduction

Malignant hyperthermia (MH) is an inherited disorder which predisposes sufferers to skeletal muscle hypercontraction, severe metabolic acidosis and a potentially lethal hyperthermia (Britt, 1985; McGrath, 1986; Sessler, 1986; O'Brien, 1987; Gronert, Mott & Lee, 1988; Harriman, 1988; Heffron, 1988; Rosenberg, 1988). In man, MH is usually triggered in susceptible individuals by halothane anaesthesia. A similar condition occurring in pigs is called the porcine stress syndrome (PSS). In response to halothane such PSS-susceptible pigs quickly develop a tachycardia and hyperventilate. Cyanotic areas then form on the skin, muscles hypercontract, limb rigidity occurs and a fatal hyperthermia ensues. Postmortem reveals oedematous, malodourous muscle with a severely disrupted structure. This meat is termed pale soft exudative (PSE) and cannot be used commercially. As well as halothane, stresses caused by transportation, exercise, feeding, mating and parturition will cause the development of a MH attack in pigs (Mitchell & Heffron, 1982).

PSS is regarded as a good model for human MH (Gronert, 1980) and has been used for much of the research into the mechanisms and causes of the disease. There is no uncomplicated and reliable diagnostic test for MH/PSS. Induction of limb rigidity by halothane inhalation will identify PSS-susceptible individuals and cause only limited mortality when performed by a skilled operator (Webb, 1980). In humans the disease is often recognised in a family when one person undergoes halothane anaesthesia prior to a surgical procedure. Thereafter muscle samples from other members of the family can be subjected to *in vitro* contracture tests to identify abnormal reaction to either halothane or caffeine. In view of its fatal consequences in humans and economic losses to the swine industry, particularly with strains of pigs which have been bred for leanness and fast growth, much research has been directed towards understanding the mechanism(s) which cause MH.

Comparison of MH and PSS

Similarities between MH and PSS include: induction of muscle rigidity and hyperthermia by halothane (Gronert, 1980; Britt, 1985), a generalised cell membrane defect (Britt, 1985), and an abnormal Ca^{2+} homeostasis in white blood cells and muscle tissue (Nelson, 1988). Despite the similar phenotypes of PSS and MH, PSS is passed from generation to generation by a recessive gene whereas MH in humans is dependent on a dominant gene, probably with variable penetration (MacLennan *et al.*, 1990). Another difference is that histological abnormalities are often demonstrated in MH-susceptible humans but rarely in PSS-susceptible pigs before a fatal stress attack. This may be due to pigs being marketed at an age before changes have developed (Gronert, 1980). PSS frequently occurs in the absence of halothane, provoked by stresses associated with normal management of the pigs. In man MH rarely occurs in the absence of a pharmacological stress (O'Brien *et al.*, 1990). The induction of PSS by stress can be a problem when testing anaesthetics for ability to induce MH. For example, in one study up to 18% pigs had a stress attack when given oxygen via a facemask (McGrath, 1986).

Enzyme/protein abnormalities in MH/PSS

MH/PSS have been associated with decreased glutathione peroxidase, glucose-6-phosphate dehydrogenase, adenylate kinase or myophosphorylase B activities or increased acetylcholinesterase and adenylate cyclase activities and increased conversion of phosphorylase b to phosphorylase a. These differences are not observed consistently (Table 1), hence none can be unequivocally associated with MH/PSS. A major problem when looking for enzyme or other abnormalities in MH/PSS is the choice of a suitable control. For measurements made with MH-susceptible humans, ideally the controls should be a group of age/sex matched individuals. In investigations of PSS the control tissue should come from age matched animals of the same breed and genetic pool. Furthermore when enzymes, such as glutathione peroxidase, whose activity is dependent upon the supply of dietary micronutrients are determined, it is essential that the subjects and controls have been maintained on the same diet at least from weaning. Lack of attention to such detail and decreased stability of tissues from MH/PSS-susceptible individuals are probably major reasons for the differences in enzyme activities which have been proposed, at one time or another, to be involved in the pathogenesis of MH.

Table 1. *Apparent enzyme abnormalities in MH/PSS*

Enzyme	Species	Activity inMH/ PSS	Reference
Glutathione peroxidase	pig	decrease	Schanus *et al.* (1981)
	dog	no change	O'Brien *et al.* (1984)
	pig	no change	Duthie *et al.* (1989*b*)
Glucose 6 phosphate dehydrogenase	pig	decrease	Schanus *et al.* (1981)
	pig	no change	Duthie & Arthur (1987)
	dog	small decrease	O'Brien *et al.* (1984)
	human	no change	Schanus *et al.* (1982)
	human	small decrease	Younker, De Vore & Hartlage (1984)
Acetylcholinesterase	pig	increase	Mickelson *et al.* (1987*b*)
Adenylate kinase	human	decrease	Schmitt, Schmidt & Ritter (1974)
	pig	no change	Marjanen & Denborough (1982) Marjanen, Shaw & Denborough (1983)
Adenylate cyclase	human	increase	Willner, Cerri & Wood (1981)
	human	increase	Ellis *et al.* (1984)
	human	plasma cAMP increase	Stanec & Stefano (1984)
	pig	no change	Sim, White & Denborough (1987)
Myophosphorylase B	human	decrease	Isaacs, Badenhorst & Du Sautoy (1989)
Pa to Pb conversion [a]	human	increase	Ellis *et al.* (1984)

Notes:
[a]Pa, Phosphorylase a; Pb. Phosphorylase b

Mechanisms

Since MH can cause muscle rigidity and heat production by the body much research has concentrated on possible defects in the systems for regulating Ca^{2+} in muscle. The rationale being that increases in free Ca^{2+}

concentration in myoplasm and mitochondria would cause all the symptoms of MH or PSS (for reviews see O'Brien, 1987; Nelson, 1988). Halothane is often used to induce changes in Ca^{2+} metabolism in muscle preparations from MH- or PSS-susceptible individuals and normal controls. This type of approach does not take into account that the condition can be induced by many forms of stress as well as pharmacologically. Hence proposed mechanisms for the pathogenesis of MH/PSS do not always offer an explanation as to how the variety of stresses cause one common final disorder. The possibility therefore exists that changes in cell Ca^{2+} in MH/PSS derive from a general cell defect which is exacerbated by pharmacological stress induced by halothane or metabolic changes associated with physiological activities such as exercise.

Reactive free radicals are produced as a consequence of oxidative cell metabolism as well as by metabolism of halothane in the endoplasmic reticulum (Poyer *et al.*, 1981; Plummer *et al.*, 1982; Halliwell & Gutteridge, 1989). Thus a mechanism involving free-radical formation as one of the primary events in the pathway leading to MH is a plausible method of explaining how diverse stimuli can cause a common pathological endpoint. This review will consider the changes in Ca^{2+} metabolism demonstrated in MH/PSS and how these changes might result from a generalised membrane defect which is exacerbated by physiologically- or pharmacologically-induced increases in free-radical formation.

Free radicals and tissue damage

Free radicals are discussed briefly in this section and the reader is referred to more detailed descriptions of radical chemistry and biochemistry in this volume. Generally, biological molecules contain stable paired electrons in chemical bonds, whereas free radicals are defined as species which are capable of independent existence and contain one or more unpaired electrons. Hence free radicals are chemically reactive in trying to form more stable compounds with paired electrons in double bonds. In forming more stable compounds free radicals often 'extract' electrons from other previously stable molecules thus generating another free radical. This ability of one free radical to generate another free radical is the basis of some chain reactions. Lipid peroxidation is one such chain reaction and could be the cause of cell membrane damage and injury in some diseases (Halliwell & Gutteridge, 1989). The major initiator of lipid peroxidation in the cell is probably the hydroxyl radical or a closely related radical species of very similar structure and reactivity (Halliwell & Gutteridge, 1989). The major source of hydroxyl radicals is thought to be metabolism involving oxygen. Superoxide ($O_2^{\cdot-}$) is formed by single-electron reduction of oxygen; the electrons leak from the carriers of the respiratory chain and react with the oxygen (Foreman & Boveris, 1982). Superoxide is metabolised via superoxide dismutases to hydrogen peroxide. There are two types of superoxide dismutase in mammalian cells: a Cu/Zn-containing form in the cytoplasm and a manganese-containing form in the mitochondria. Hydrogen peroxide formed by the superoxide dismutases is further metabolised by the selenium-containing glutathione peroxidase in the cytosol and plasma. Glutathione peroxidase will use glutathione to chemically reduce hydrogen peroxide and a wide range of lipid hydroperoxides. However, it will not metabolise fatty acid hydroperoxides when they are attached to phospholipids. A selenium-containing phospholipid hydroperoxide peroxidase, which will fulfil this function, has, however, been discovered (Ursini, Maiorini & Gregolin, 1985). The combined effect of the superoxide dismutases and the glutathione peroxidases is thought to be the prevention of a transition-metal-catalysed Haber–Weiss reaction which could result in the formation of hydroxyl radicals (Halliwell & Gutteridge, 1989).

Despite the actions of glutathione peroxidase and superoxide dismutase free radicals may still form in the cell and attack lipid membranes. In membranes the major chain-breaking free-radical scavenger is vitamin E, which occurs almost exclusively as the α-tocopherol isomer. Vitamin E donates a hydrogen to lipid radicals to stop peroxidation; the resultant tocopheryl radical is not sufficiently reactive to abstract a hydrogen from membrane fatty acids and restart peroxidation. Thus in preventing peroxidation tocopheryl radicals would build up in the cell membrane if the vitamin E was not regenerated. In *in vitro* systems vitamin C and glutathione react with the tocopherol radical to regenerate vitamin E. For these mechanisms to operate *in vivo* these water-soluble molecules would probably have to react with the tocopherol radical at the surface of the lipid cell membrane.

MH and PSS; vitamin E deficiency and free-radical activity

Deficiency in many of the cell's antioxidant systems may result in disease. Animals which consume inadequate amounts of vitamin E and selenium have low membrane vitamin E concentrations and low tissue glutathione peroxidase activities. In rats and pigs this can result in liver necrosis and in ruminants and pigs skeletal muscle myopathies and cardio-myopathies, diseases thought to be associated with peroxidative free-radical-mediated damage occurring in lipid components of cell membranes (reviewed by Combs & Combs, 1986). Despite PSS-susceptible pigs having normal or slightly increased tissue vitamin E concentrations (Duthie *et al.*, 1987*b*), there are some features of PSS which are similar to vitamin E deficiency and are consistent with both conditions being caused, at least in part, by free radical mediated damage to cell membranes (Duthie & Arthur, 1989).

1. Both vitamin E-deficient animals and PSS pigs have increased activity of the muscle enzymes creatine kinase and pyruvate kinase in plasma, indicative of damage to muscle cell membranes (Duthie & Arthur, 1987; Duthie, Arthur & Mills, 1987*a*; Duthie *et al.*, 1988*b*).
2. Erythrocytes from vitamin E-deficient animals and from some PSS pigs show increased susceptibility to osmotically induced lysis. This effect is not, however, a consistent feature of PSS (Duthie *et al.*, 1989*a*).
3. Polyunsaturated fatty acids of the cell are susceptible to peroxidation particularly in vitamin E deficiency. Consequently plasma from vitamin E-deficient animals has increased concentrations of thiobarbituric acid reactive substances (TBARS) and conjugated dienes, both products formed during lipid peroxidation. Increases in TBARS and conjugated dienes are also detected when plasma from PSS-susceptible pigs is compared with plasma from stress-resistant pigs of the same breed (Duthie, Arthur & Hoppe, 1988*a*; Duthie *et al.*, 1989*b*).
4. Compared with samples from vitamin E-supplemented animals tissue homogenates from vitamin E-deficient animals produce more of the hydrocarbon gases ethane and pentane, evidence of increased free-radical-mediated lipid peroxidation. Similarly incubated muscle homogenates from PSS pigs produce more pentane than do muscle homogenates from stress resistant pigs. Incubation of red blood cells from vitamin E-deficient animals or PSS-susceptible pigs with hydrogen peroxide results in increased TBARS formation compared with that found in normal animals (Duthie *et al.*, 1989*b*; Halliwell & Gutteridge, 1989).
5. Abnormalities in Ca^{2+} metabolism have been demonstrated in MH/PSS (see Gronert, 1980) and vitamin E may influence membrane permeability to Ca^{2+} (Jackson, Jones & Edwards, 1985).

Other observations, not all directly related to vitamin E-deficiency, support a role for free radicals in MH/PSS. Halothane ($CF_3CHBrCl$)

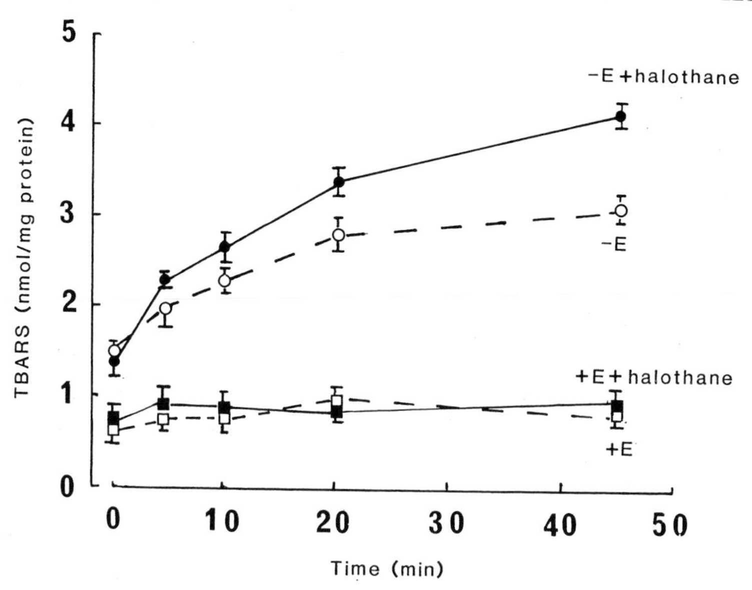

Fig. 1. TBARS production in incubations of liver microsomes from vitamin E-deficient and -sufficient rats. Liver microsomes were prepared from rats which had consumed vitamin E-deficient or vitamin E-sufficient diets for 8 weeks (Duthie *et al.*, 1987*a*). TBARS production was determined in incubations which contained 0 or 10 μM halothane.

triggers the MH response in susceptible individuals and can form free radicals *in vivo* (Forni *et al.*, 1983; Poyer *et al.*, 1981; Plummer *et al.*, 1982). As halothane exacerbates the enhanced lipid peroxidation of microsomal preparations from vitamin E-deficient rats (Fig. 1), halothane-derived radicals could be an initiating factor for the MH attack. Moreover, incubation of liver homogenates from PSS-susceptible pigs produces more TBARS than homogenates than normal pigs. This TBARS formation is accompanied by increased free-radical production demonstrated by spin trapping and ESR spectroscopy (Duthie *et al.*, 1990). The rapid development of the PSS attack is consistent with the rapid chain reaction of free-radical-mediated lipid peroxidation. In contrast to the reported decreased glutathione peroxidase activity in PSS and MH (Schanus *et al.*, 1981; 1982), when PSS-susceptible pigs are compared with normal individuals of the same breed consuming identical diets, muscle glutathione peroxidase activity is increased (Duthie *et al.*, 1989*b*). The enhanced glutathione peroxidase activity may be an indication of excessive free-radical activity in the PSS-susceptible pigs. Blood cells and muscle from PSS-susceptible pigs also have higher total GSH concentrations than tissue from normal animals (Duthie *et al.*, 1989*a*; Duthie & Arthur, 1989). Such increases in GSH occur in animals undergoing an oxidant stress or in animals with a deficiency in cell antioxidant systems (Allen *et al.*, 1988; Rotruck *et al.*, 1972). The similarities between PSS and vitamin E deficiency and the indications of free-radical activity associated with PSS led to investigations of the effects of supplementing PSS-susceptible pigs with the vitamin E.

Vitamin E supplementation of the PSS-susceptible pig and evidence for membrane abnormalities in MH/PSS

Supplementation of diets with 235 IU vitamin E/kg increases plasma vitamin E concentrations in both PSS-susceptible and normal pigs when compared with animals consuming the basal diet containing 10 IU vitamin E/kg (Fig. 2). Since no difference was apparent in plasma vitamin E concentrations between PSS-susceptible and normal pigs with similar dietary vitamin E intakes, the PSS-susceptible pigs were not vitamin E deficient. Before vitamin E supplementation plasma creatine kinase and pyruvate kinase activities were higher in the PSS-susceptible pigs than the normal pigs. After 5 weeks the enzyme activities in the vitamin E-supplemented PSS-susceptible pigs were lower than in the PSS-susceptible

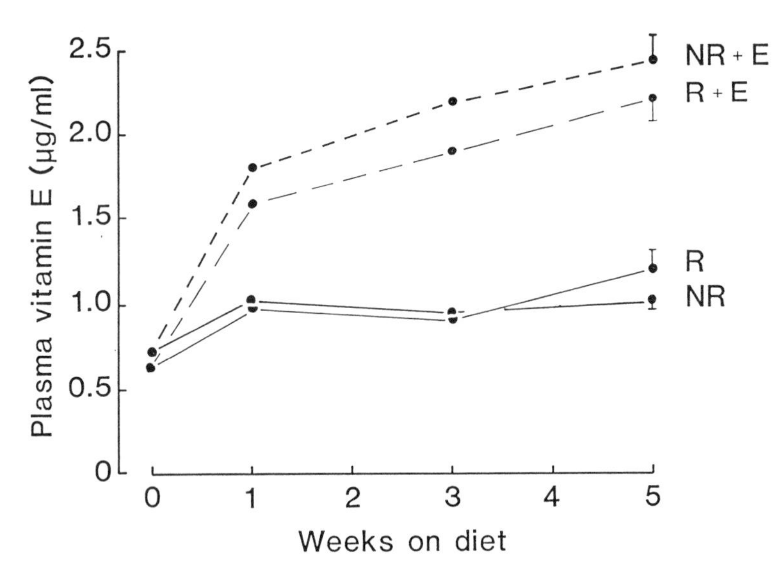

Fig. 2. Plasma vitamin E concentrations in PSS-susceptible (R) and PSS-resistant pigs (NR). Pigs (8/group; 8–10 weeks old) consumed diets containing 10 or 235 IU vitamin E/ kg (as α-tocopheryl acetate, +E).

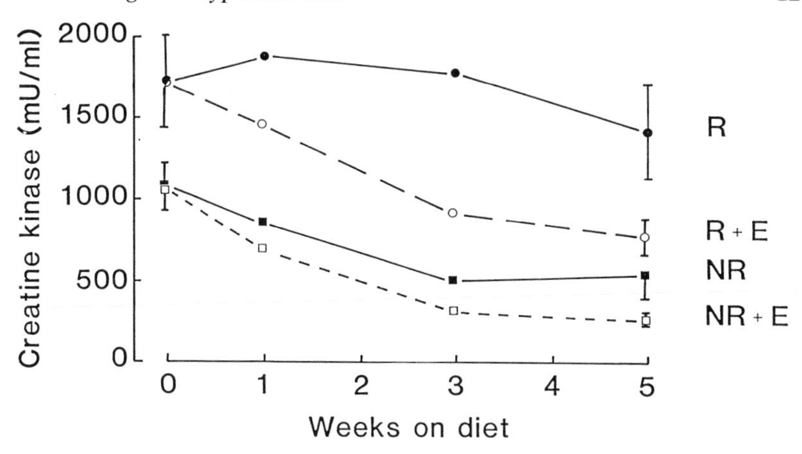

Fig. 3. Plasma creatine kinase activities in PSS-susceptible (R) and PSS-resistant pigs (NR). Experimental animals are described in Fig. 2.

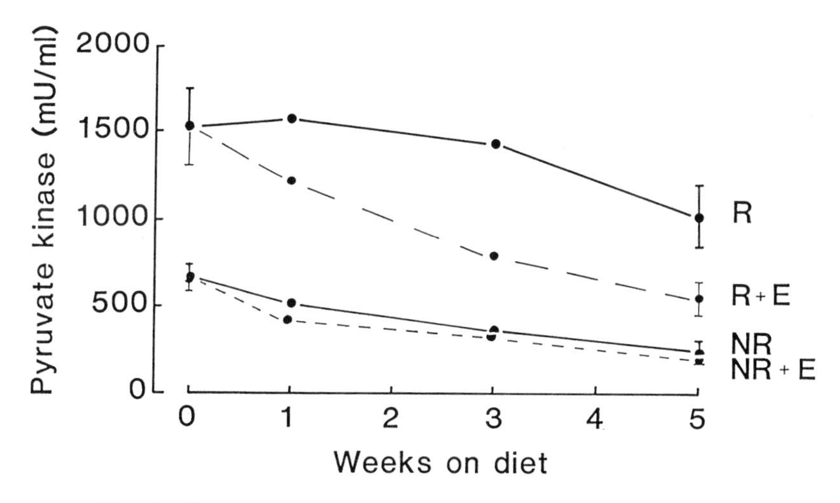

Fig. 4. Plasma pyruvate kinase activities in PSS-susceptible (R) and PSS-resistant pigs (NR). Experimental animals are described in Fig. 2.

pigs consuming the lower amount of vitamin E (Figs 3 and 4). Vitamin E supplementation of PSS-susceptible pigs also decreases plasma TBARS concentrations, prevents hydrogen-peroxide-induced TBARS formation in incubations of blood cells and decreases pentane production by muscle homogenates (Duthie *et al.*, 1989*b*; Hoppe *et al.*, 1989). The effects of vitamin E supplementation are consistent with a counteraction of an antioxidant abnormality or membrane defect in the PSS-susceptible pig.

Spin labelling and electron paramagnetic resonance spectroscopy indicate that red cell membranes from PSS- and MH-susceptible individuals and sarcoplasmic reticulum membranes from PSS-susceptible pigs are disordered to a greater degree by halothane than membranes from normal subjects (Ohnishi & Ohnishi, 1988; Ohnishi *et al.*, 1988). There is also evidence for membrane defects in heart muscle, smooth muscle, motor nerves, brain cells, platelets, lymphocytes and Islets of Langerhans of MH/PSS-susceptible individuals (Britt, 1985; Basrur, Bouvet & McDonell, 1988). A membrane defect would explain why lymphocytes from MH/PSS-susceptible individuals will accumulate Ca^{2+} when incubated in the presence of halothane, whereas cells from normal subjects will not (Klip *et al.*, 1986; 1987).

Calcium free radicals and lipid peroxidation

Oxidant stress increases the normally very low free Ca^{2+} concentrations in the cell. This may depend on damage to the Ca-ATPases of the plasma membrane and endoplasmic reticulum. These enzymes normally remove Ca^{2+} from the cytoplasm and contain essential thiol groups which are particularly sensitive to oxidation which causes enzyme inactivation (Kagan *et al.*, 1989). Lipid peroxidation, initiated by Fe^{2+}, can be further stimulated by as little as 0.1 μM Ca^{2+} (Braughler, 1988). Hence lipid peroxidation occurring at the early stages of an MH/PSS attack could disrupt Ca^{2+} homeostasis causing increased myoplasmic/cytoplasmic Ca^{2+} concentrations sufficient to result in muscle hypercontraction and the metabolic hyperthermia.

Calcium homeostasis in the muscle cell

Contraction of skeletal muscle is initiated by rapid increases in myoplasmic Ca^{2+} ions which bind to troponin C, a protein of the thin filament. In addition to the excitation–contraction coupling involved in muscle contraction, the regulation of changes in cytosolic free calcium concentrations is crucial to the mediation of many other cellular processes such as the control of enzyme activities, signal transduction and messenger–hormonal interactions (Cheung *et al.*, 1986). The total cell calcium content ranges from 0.1 to 1.0 mmol/kg whereas techniques such as Quin-2 fluorescence, calcium-selective microelectrodes and nuclear magnetic resonance confirm that resting cytosolic free Ca^{2+} concentrations are in the order of only 100 nM (Baker, 1986). Consequently, 99.9% of cell calcium is either bound or sequestered in subcellular organelles such as endoplasmic and sarcoplasmic reticulum and mitochondria. Appreciable amounts of calcium are bound to the phospholipid and glycoprotein components of plasma membranes, intracellular phosphates, and to cytosolic proteins such as calmodulin and troponin C (Cheung *et al.*, 1986: Hertzberg, Moult & James, 1986).

The Ca^{2+} concentration in the extracellular fluid is three to four orders of magnitude higher than in the cytoplasm. Although there is a small and steady Ca^{2+} influx into the cell a large electrochemical gradient is maintained by a low cell permeability to calcium and a variety of mechanisms which actively remove calcium from the myoplasm of the relaxed muscle. Such mechanisms also serve to ensure subsequent relaxation of contracted muscle by rapidly restoring resting myoplasmic Ca^{2+} concentrations after the excitation event and are summarised in Fig. 5.

The Ca-ATPases located on the sarcolemma and sarcoplasmic reticulum are membrane-bound proteins which remove two Ca^{2+} from the sarcoplasm for each mole of ATP hydrolysed in order to bring about muscle relaxation (Jencks, 1989). Although magnesium ions are required for the operation of the Ca-ATPase pump, Mg^{2+} is not transported in exchange for Ca^{2+} and the variety of steps in the reaction cycle of the pump are fully reversible (Green *et al.*, 1986). A reversible Na^{+}/Ca^{2+} antiporter located in the plasma membrane can also contribute to the efflux of calcium from the cell. Under resting conditions, Ca^{2+} inflow and outflow through the exchanger are roughly in balance. Ca^{2+} entry via the exchanger is enhanced by plasmalemma depolarisation and/or by a rise in intracellular Na^{+} whereas efflux results from repolarisaton and/or a rise in internal free Ca^{2+}. The exchanger has a lower affinity but larger transport capacity for Ca^{2+} than the Ca-ATPase pump. Consequently, following a period when cytoplasmic Ca^{2+} is elevated, the antiporter may operate in parallel with the Ca-ATPase in removing Ca^{2+} from the cell. The former effects a swift reduction in free Ca^{2+} towards the resting level and the action of the pump establishes a lower concentration than can be achieved by the exchanger alone (Baker, 1986).

Release of calcium from the sarcoplasmic reticulum occurs via a channel in the ryanodine receptor, a 400 K tetrameric protein which was previously described as spanning protein and which forms junctions known as feet between sarcolemmal invaginations called transverse tubules and large sacs of sarcoplasmic reticulum referred to as terminal cisternae (Nelson, 1988; Mickelson *et al.*, 1988). In excitation–contraction coupling, acetylcholine released from the nerve ending binds to the muscle endplate. The subsequent action potential causes depolarisation of the transverse tubule membrane which is detected by a voltage-dependent dihydropyridine-sensitive receptor in close contact to the ryanodine receptor. The signal is then believed to be modulated by the ryanodine receptor, resulting in the opening of a channel within the core

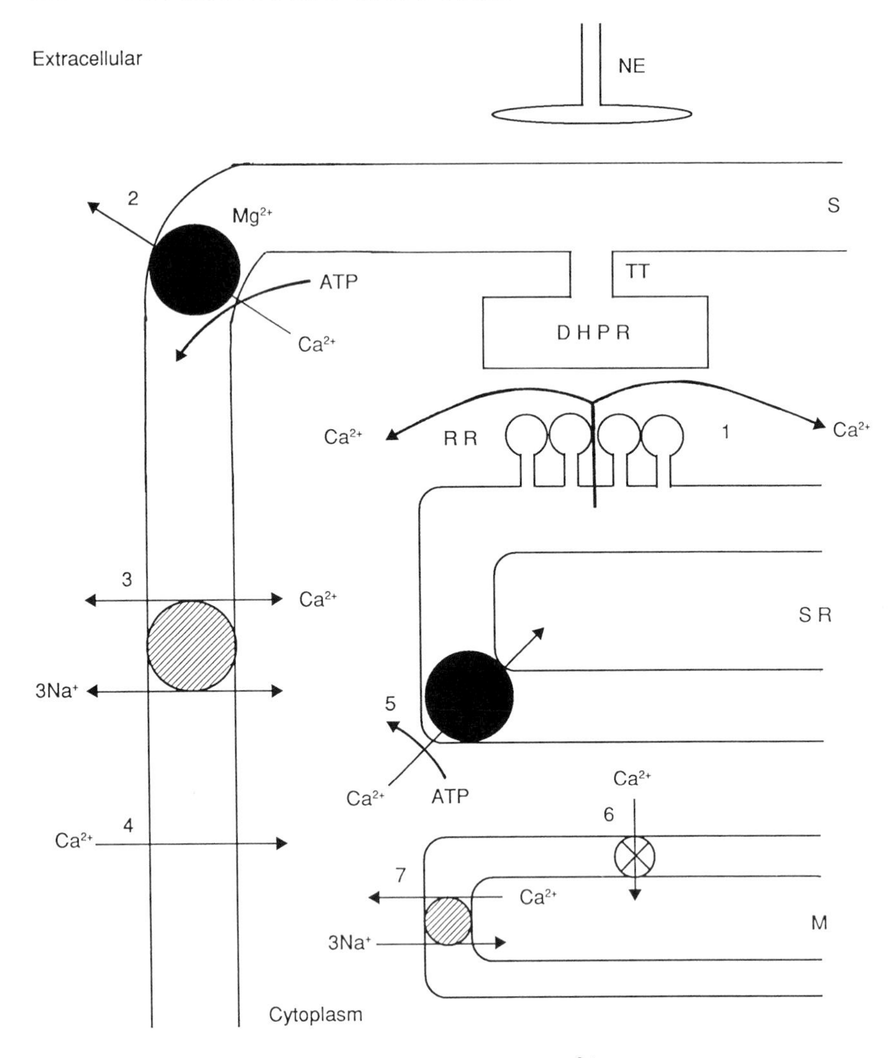

Fig. 5. Major mechanisms of cytosolic Ca^{2+} regulation in the muscle cell. 1. Voltage dependant. 2. Sarcolemmal Ca^{2+}-ATPase. 3. Na^{+}/Ca^{2+} antiporter. 4. Steady-state Ca^{2+} influx. 5. Sarcoplasmic reticulum Ca^{2+}-ATPase. 6. Mitochondrial uniporter. 7. Na^{+}/Ca^{2+} antiporter. Abbreviations: S, sarcolemma; SR, sarcoplasmic reticulum; M, mitochondrion; NE, nerve ending; TT, transverse tubule; DHPR, voltage dependent dihydropyridine receptor; RR, ryanodine receptor.

of the tetrameric structure (Gill, 1989). Channel opening also appears to be dependent on small priming doses of Ca^{2+}. Calsequestrin, a 46 K protein primarily associated with the terminal cisternae of the sarcoplasmic reticulum may provide such a source of Ca^{2+} for release into the myoplasm to initiate contraction (Nelson, 1988). In non-muscle tissues, inositol 1,4,5-triphosphate and associated receptor (Berridge, 1986; Berridge & Irvine, 1989) perform an analogous function to the voltage-dependent dihydropyridine-sensitive receptor and ryanodine receptor system in muscle (Gill, 1989).

Mitochondria are unlikely to be important in the fine tuning of cytosolic Ca^{2+} (Carafoli, 1986). The endogenous mitochondrial Ca^{2+} pool in muscle amounts to only 1 nmol/mg protein and the amount that can be contributed to the cytosol is only of the order of 50 nmol/g wet tissue (Nicholls, 1986). Therefore, it is unlikely that Ca^{2+} efflux mediated by the Na^{+}/Ca^{2+} antiporter on the inner mitochondrial membrane is of sufficient magnitude to trigger muscle contraction. Linkage between mitochondrial respiration and ATP synthesis provides the driving force for a uniporter on the inner membrane to sequester Ca^{2+} from the cytosol to the mitochondrial matrix. The activities of the two transporting mechanisms function to control mitochondrial Ca^{2+} concentrations which regulate the dehydrogenase enzymes involved in the initiation and operation of the Krebs Cycle. Although muscle mitochondrial Ca^{2+} appears to operate in a dehydrogenase regulating mode rather than in a cytoplasmic buffering mode, pathologically high increases in cytoplasmic calcium could result in increased influx to the mitochondrial matrix and result in the uncoupling of oxidative phosphorylation (Britt, 1985), and the subsequent generation of heat. Some studies (Cheah & Cheah, 1973; 1981; Cheah, 1984: Fletcher & Rosenberg, 1986) suggest that the activity of mitochondrial Ca^{2+}-activated phospholipase A_2 is elevated in MHS muscle. Resulting release of free fatty acids may then increase release of Ca^{2+} from the sarcoplasmic reticulum. The Ca^{2+} channels on sarcoplasmic reticulum therefore could be rendered superpermeable by a basic mitochondrial abnormality. Both enzymic and non-enzymic oxidants induce release of Ca^{2+} from mitochondria. Subsequent excessive Ca^{2+} cycling by mitochondria leads to uncoupling of oxidative phosphorylation and disruption of the cell's ATP supply. As a result, the function of sarcolemmal and sarcoplasmic reticulum Ca^{2+}-ATPases is impaired leading to uncontrolled rises in cytosolic Ca^{2+} (Richter & Frei, 1988) which could result in muscle contraction.

Due to the limb rigidity that occurs during an MH episode, at one time or another abnormalities in most of the Ca^{2+} regulatory mechanisms have been proposed to be the cause of or involved in the etiology of the MH/

PSS. However, specific defects in any particular mechanism have yet to be consistently demonstrated.

Proposed defects in Ca^{2+} regulation in MH/PSS

Increased myoplasmic Ca^{2+} concentrations have been detected in MH/PSS using microelectrodes inserted in muscle of anaesthetised MH/PSS-susceptible individuals. Halothane treatment does not increase Ca^{2+} concentrations in normal muscle but rapid increases occur on treatment of MH/PSS muscle (Lopez *et al.*, 1985; 1986; 1988).

Suggestions that the sarcolemmal Ca^{2+} regulatory mechanisms are abnormal are confounded by the difficulty in discerning whether the defect lies within the membrane itself or resides within the structures of the sarcoplasmic reticulum associated with the sarcolemma (Fig. 5; Gronert *et al.*, 1988). Moreover, studies which appear to show increased plasmalemmal depolarisation, enhanced Ca^{2+} permeability of transverse tubules or abnormal sarcolemmal Ca-ATPase activity and lipid profiles (Gallant *et al.*, 1979; 1982; Niebroj-Dobosz, Kwiatkowski & Mayner-Zawadzka, 1984; Mickelson *et al.*, 1987*a*; Rock & Kozak-Reiss, 1987) may merely reflect variation and contamination during tissue preparation (Fletcher *et al.*, 1988; Gallant, 1988; Iaizzo *et al.*, 1988; Ervasti, Mickelson & Louis, 1989).

Abnormalities or increased sensitivity to halothane of Ca-ATPase have been described in muscle membranes of MH/PSS-susceptible individuals. The effects are consistent with the ability of halothane to form reactive free radicals which would inactivate the Ca-ATPase by oxidising essential thiol groups. However, several publications have concluded that there was no good evidence for abnormal Ca-ATPase function in sarcoplasmic reticulum Ca^{2+} uptake in MH/PSS (for review see Nelson, 1988). Reported defects in Ca-ATPase could therefore be secondary consequences of tissue preparation or free-radical activity during the stress attack.

There have been consistent demonstrations of a defect in the mechanism or control of Ca^{2+} release from the sarcoplasmic reticulum of muscle from MH/PSS subjects is more sensitive than control preparations to stimulation by halothane, caffeine and low priming doses of Ca^{2+} (Ohnishi *et al.*, 1986). The changes in sarcoplasmic reticulum Ca^{2+} release have been attributed to abnormalities in Ca^{2+}-sensitive Ca^{2+} channels such as the voltage-sensitive ryanodine receptor (Mickelson *et al.*, 1988; Knudson *et al.*, 1990). The receptor from muscle of PSS-susceptible pigs has altered ryanodine binding compared with normal, consistent with an open configuration allowing Ca^{2+} release from the sarcoplasmic reticulum. The involvement of the ryanodine receptor in MH/PSS has received further support from genetic linkage and mapping studies of MH-susceptible families. In families in which MH is inherited as an autosomal dominant trait the MH locus is on chromosome 19q12–13.2, which is linked to DNA markers for glucose phosphate isomerase (McCarthy *et al.*, 1990). In PSS-susceptible pigs the locus for glucose phosphate isomerase is genetically linked to that for PSS (Andresson & Jensen, 1977). Additional linkage studies show that in MH-susceptible families the MH phenotype segregates with markers of the ryanodine receptor in chromosome 19q (MacLennan *et al.*, 1990). These studies strongly support a primary role for abnormalities in the ryanodine receptor or in the ryanodine gene in the changes in Ca^{2+} release associated with MH/PSS. However, changes in the ryanodine receptor may not explain increases in free Ca^{2+} observed after halothane treatment of blood lymphocytes from PSS-susceptible pigs and from MH-susceptible patients (Klip *et al.*, 1986; 1987).

Conclusions

Several studies show that the pathogenesis of MH/PSS can be related: (1) to a cell membrane abnormality or (2) to changes in the regulation of cell Ca^{2+} concentration or (3) to a defect in cell antioxidant systems. Fig. 6 shows a hypothesis for the integration of these observations. After stress or halothane anaesthesia the MH/PSS-susceptible individual may not be able to cope with increased free-radical formation due to an antioxidant abnormality (Duthie & Arthur, 1989). Increased free-radical formation could then overwhelm the compromised antioxidant system to initiate chain reactions causing damage to the lipids and proteins in cell membranes. Damage to muscle cell membranes after halothane anaesthesia of PSS-susceptible pigs would account for the increased plasma pyruvate kinase and creatine kinase observed after treatment. An inherent chronic membrane defect would also allow release of the enzymes.

A further effect of free-radical activity would be to impair the activity of the proteins involved in regulation of cell Ca^{2+} ion concentrations. This may cause or amplify an abnormality in the sarcoplasmic ryanodine receptor and associated sarcoplasmic reticulum Ca^{2+} channel. The ryanodine receptor abnormality would also result from a membrane defect affecting protein conformation or derive directly from a defect in the ryanodine gene, which is located very close to the gene for MH/PSS (MacLennan *et al.*, 1990; MacCarthy *et al.*, 1990). Thus antioxidant- or membrane- or Ca^{2+}-homeostasis related abnormalities may all cause increased myoplasmic Ca^{2+} ion concentrations in MH/PSS-susceptible

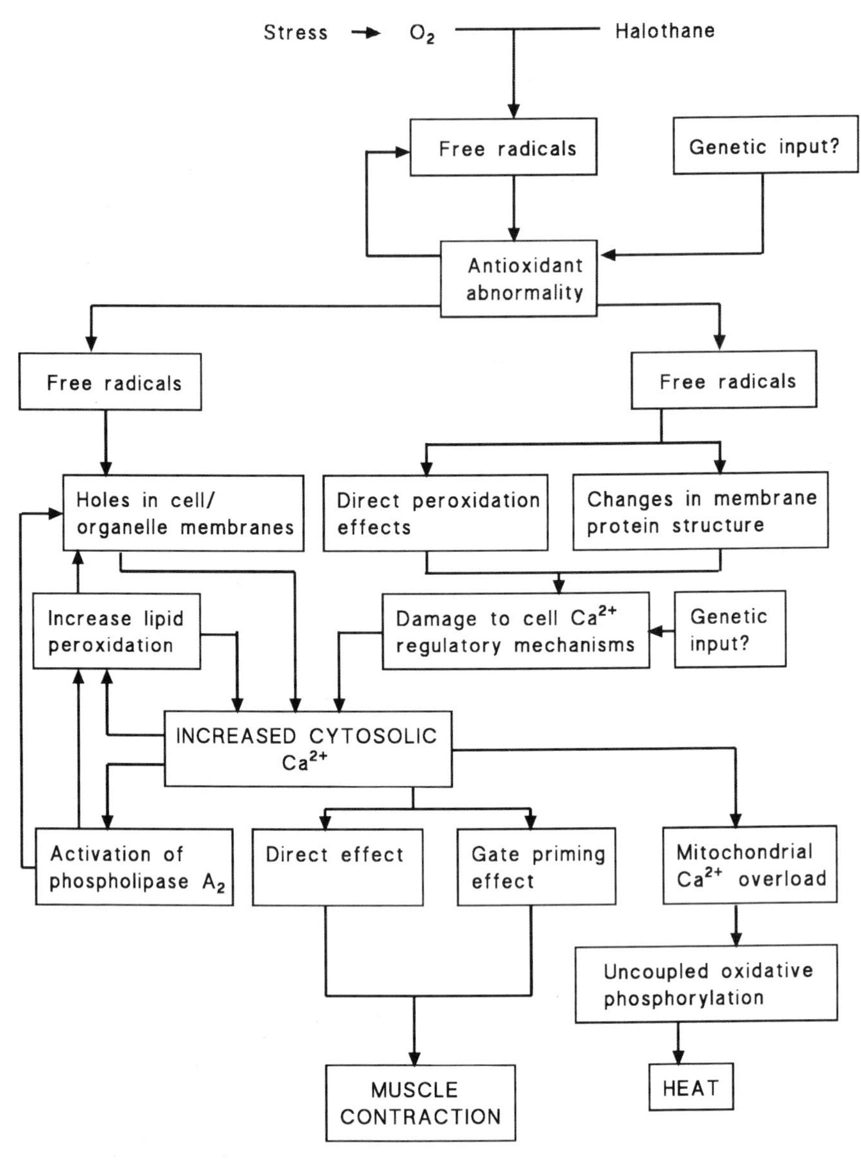

Fig. 6. Scheme for the integration of the various mechanisms proposed for the pathogenesis of MH/PSS.

individuals. The muscle contraction and heat production characteristic of MH/PSS would then result from the elevated Ca^{2+} concentrations.

Although a genetically induced abnormality in the ryanodine receptor would explain many of the changes in muscle associated with MH/PSS, the alterations in other organs and red and white blood cells are less easy to relate to a muscle-specific mechanism for Ca^{2+} regulation. Antioxidant or membrane abnormalities, which may be related to free-radical activity, are therefore possible causes of general tissue effects of MH/PSS including that on the ryanodine receptor. Thus further research is required to identify finally the abnormality which can explain all the symptoms and biochemical and physiological consequences of MH/PSS.

References

Allen, K.G.D., Arthur, J.R., Morrice, P.C., Nicol, F. & Mills, C.F. (1988). Copper deficiency and tissue glutathione concentrations in the rat. *Proceedings of the Society for Experimental Biology and Medicine* **187**, 38–43.

Andresson, E. & Jensen, P. (1977). Close linkage established between the HAL locus for halothane sensitivity and the PHI (phosphohexose isomerase) locus in pigs of the Danish Landrace breed. *Nordic Veterinary Medicine* **29**, 502–4.

Baker, P.F. (1986). Introduction. In *Calcium and the Cell* (*Ciba Foundation Symposium 122*), ed. D. Evered & J. Whelan, pp. 1–4. Chichester: Wiley.

Basrur, P.K., Bouvet, A. & McDonell, W.N. (1988). Open canicular system of platelets in porcine stress syndrome. *Canadian Journal of Veterinary Research* **52**, 380–5.

Berridge, M.J. (1986). Inositol triphosphate and calcium mobilisation. In *Calcium and the Cell* (*CIBA Foundation Symposium 122*), ed. D. Evered & J. Whelan, pp. 39–57. Chichester: Wiley.

Berridge, M.J. & Irvine (1989). Inositol phosphates and cell signalling. *Nature* **341**, 197–205.

Braughler, J.M. (1988). Calcium and lipid peroxidation. In *Oxygen Radicals and Tissue Injury*, ed. B. Halliwell, pp. 99–104. Maryland: FASEB.

Britt, B.A. (1985). Malignant Hyperthermia. *Canadian Anaesthetists' Society Journal* **32**, 666–77.

Carafoli, E. (1986). General discussion (Aspects of regulation of calcium in the cell). In *Calcium and the Cell* (*CIBA Foundation Symposium 122*), ed. D. Evered & J. Whelan, pp. 115–19. Chichester: Wiley.

Cheah, K.S. (1984). Skeletal-muscle mitochondria and phospholipase A_2 in malignant hyperthermia. *Biochemical Society Transactions* **12**, 358–60.

Cheah, K.S. & Cheah, A.M. (1973). Comparative studies of the mitochondrial properties of *longissimus dorsi* muscles of Pietran and Large white pigs. *Journal of the Science of Food and Agriculture* **24**, 51–61.

Cheah, K.S. & Cheah, A.M. (1981). Skeletal muscle mitochondrial phospholipase A_2 and the interaction of mitochondria and sarcoplasmic reticulum in porcine malignant hyperthermia. *Biochimica et Biophysica Acta* **638**, 40–9.

Cheung, J.Y., Bonventre, J.V., Malis, C.D. & Leaf, A. (1986). Calcium and ischaemic injury. *The New England Journal of Medicine* **314**, 1670–6.

Combs, G.F. & Combs, S.B. (1986). *The Role of Selenium in Nutrition*, pp. 532. New York: Academic Press.

Duthie, G.G. & Arthur, J.R. (1987). Blood antioxidant status and plasma pyruvate kinase activity of halothane-reacting pigs. *American Journal of Veterinary Research* **48**, 309–10.

Duthie, G.G. & Arthur, J.R. (1989). The antioxidant abnormality in the stress susceptible pig: The effects of vitamin E supplementation. *Annals of the New York Academy of Science* **570**, 322–34.

Duthie, G.G., Arthur, J., Bremner, P., Kikuchi, Y. & Nicol, F. (1989*a*). Increased lipid peroxidation of erythrocytes of stress-susceptible pigs: An improved diagnostic test for porcine stress syndrome. *American Journal of Veterinary Research* **50**, 84–9.

Duthie, G.G., Arthur, J.R. & Hoppe, P.P. (1988*a*). Porcine stress syndrome, free radicals and vitamin E. In *Oxygen Radicals in Biology and Medicine*, ed. M. Simic, pp. 605–9. New York: Plenum.

Duthie, G.G., Arthur, J.R. & Mills, C.F. (1987*a*). Tissue damage in vitamin E-deficient rats is not detected by expired ethane and pentane. *Free Radical Research Communications* **4**, 21–5.

Duthie, G.G., Arthur, J.R., Mills, C.F., Morrice, P.C. & Nicol, F. (1987*b*). Anomalous tissue vitamin E distribution in stress susceptible pigs after dietary vitamin E supplementation and effects on plasma pyruvate kinase and creatine kinase activities. *Livestock Production Science* **17**, 169–78.

Duthie, G.G., Arthur, J.R., Nicol, F. & Walker, M.J. (1989*b*). Increased indices of lipid peroxidation in stress susceptible pigs and effects of vitamin E. *Research in Veterinary Science* **46**, 226–30.

Duthie, G.G., Arthur, J., Simpson, P. & Nicol, F. (1988*b*). Plasma pyruvate kinase activity vs creatine kinase activity as an indicator of the porcine stress syndrome. *American Journal of Veterinary Research* **49**, 508–10.

Duthie, G.G., McPhail, D.B., Arthur, J.R., Goodman, B.A. & Morrice, P.C. (1990). Spin trapping of free radicals and lipid peroxidation in microsomal preparations from malignant hyperthermia susceptible pigs. *Free Radical Research Communications* **8**, 93–9.

Ellis, F.R., Halsall, P.J., Allan, P. & Hay, E. (1984). A biochemical abnormality found in muscle from unstressed malignant hyperpyrexia-susceptible muscle. *Biochemical Society Transactions* **12**, 357–8.

Ervasti, J.M., Mickelson, J.R. & Louis, C.F. (1989). Transverse tubule calcium regulation in malignant hyperthermia. *Archives of Biochemistry and Biophysics* **269**, 497–506.

Fletcher, J.E. & Rosenberg, H. (1986). In vitro muscle contractures induced by halothane and suxamethonium. *British Journal of Anaesthesia* **58**, 1433–9.

Fletcher, J.E., Rosenberg, H., Michaux, K., Cheah, K.S. & Cheah, A.M. (1988). Lipid analysis of skeletal muscle from pigs sussceptible to malignant hyperthermia. *Biochemistry and Cell Biology* **66**, 917–21.

Foreman, H.J. & Boveris, A. (1982). Superoxide and hydrogen peroxide in mitochondria. In *Free Radicals in Biology and Medicine*, vol. 5, ed. W.A. Pryor, pp. 65–90. New York: Academic Press.

Forni, L.G., Packer, J.E., Slater, T.F. & Willson, R.L. (1983). Reaction of the trichlormethyl and halothane-derived peroxy radicals with unsaturated fatty acids: a pulse radiolysis study. *Chemico-Biological Interactions* **45**, 171–7.

Gallant, E.M. (1988). Porcine malignant hyperthermia: no role for plasmalemmal depolarization. *Muscle and Nerve* **11**, 785–6.

Gallant, E.M., Godt, R.E. & Gronert, G.A. (1979). Role of plasma membrane defect of skeletal muscle in malignant hyperthermia. *Muscle and Nerve* **2**, 491–4.

Gallant, E.M., Gronert, G.A. & Taylor, S.R. (1982). Cellular membrane potentials and contractile threshhold in mammalian skeletal muscle susceptible to malignant hyperthermia. *Neuroscience Letters* **28**, 181–6.

Gill, D.L. (1989). Receptor kinships revealed. *Nature* **342**, 16–18.

Green, N.M., Taylor, W.R., Brandl, C., Korczak, B. & MacLennan, D.H. (1986). Structural and mechanistic implications of the amino acid sequence of calcium-transporting ATPases. In *Calcium and the Cell* (*CIBA Foundation Symposium 122*), ed. D. Evered & J. Whelan, pp. 93–107. Chichester: Wiley.

Gronert, G.A. (1980). Malignant hyperthermia. *Anesthesiology* **53**, 395–423.

Gronert, G.A., Mott, J. & Lee, J. (1988). Aetiology of malignant hyperthermia. *British Journal of Anaesthesia* **60**, 253–67.

Halliwell, B. & Gutteridge, J.M.C. (1989). *Free Radicals in Biology and Medicine*, 2nd edn., 543 pp. Oxford: Clarendon Press.

Harriman, D.G.F. (1988). Malignant hyperthermia myopathy – a critical review. *British Journal of Anaesthesia* **60**, 309–16.

Heffron, J.J.A. (1988). Malignant hyperthermia: biochemical aspects of the acute episode. *British Journal of Anaesthesia* **60**, 274–8.

Hertzberg, O., Moult, J. & James, M.N.G. (1986). Calcium binding to skeletal muscle troponin C and the regulation of muscle contraction. In *Calcium and the Cell* (*CIBA Foundation Symposium 122*), ed. D. Evered & J. Whelan, pp. 120–39. Chichester: Wiley.

Hoppe, P.P., Duthie, G.G., Arthur, J.R., Schoner, F.J. & Wiesche, H. (1989). Vitamin E and vitamin C supplementation and stress susceptible pigs: Effects of halothane and pharmacologically-induced muscle contractions. *Livestock Production Science* **22**, 341–50.

Iaizzo, P.A., Lehmann-Horn, F., Taylor, S.R. & Gallant, E.M. (1988). Malignant hyperthermia: effects of halothane on the surface membrane. *Muscle and Nerve* **12**, 178–83.

Isaacs, H., Badenhorst, M.E. & Du Sautoy, C. (1989). Myophosphorylase B deficiency in malignant hyperthermia. *Muscle and Nerve* **12**, 203–5.

Jackson, M.J., Jones, D.A. & Edwards, R.H.T. (1985). Vitamin E and muscle diseases. *Journal of Inherited Metabolic Disease* **8**, Suppl. 1, 84–7.

Jencks, W.P. (1989). How does a calcium pump pump calcium? *The Journal of Biological Chemistry* **264**, 18855–8.

Kagan, V.E., Bakalova, R.A., Rangelova, D.S., Stoyanovsky, D.A., Koynova, G.M. & Wolinsky, I. (1989). Oxidative stress leads to inhibition of calcium transport by sarcoplasmic reticulum in skeletal muscle. *Proceedings of the Society for Experimental Biology and Medicine* **190**, 365–8.

Klip, A., Britt, B.A., Elliott, M.E., Pegg, W., Frodis, W. & Scott, E. (1987). Anaesthetic-induced increase in ionised calcium in blood mononuclear cells from malignant hyperthermia patients. *Lancet* **i**, 463–6.

Klip, A., Britt, B.A., Elliott, M.E., Walker, D., Ramal, T. & Pegg, W. (1986). Changes in cytoplasmic calcium caused by halothane. Role of plasma membrane and intracellular Ca^{2+} stores. *Biochemistry and Cell Biology* **64**, 1181–9.

Knudson, C.M., Mickelson, J.R., Louis, C.F. & Campbell, K.P. (1990). Distinct immunopeptide maps of the sarcoplasmic reticulum Ca^{2+}-release channel in malignant hyperthermia. *Journal of Biological Chemistry* **265**, 2421–4.

Lopez, J.R., Alamo, L.A., Caputo, C., Wikinski, J. & Ledezma, D. (1985). Intracellular ionized calcium concentration in muscles from humans with malignant hyperthermia. *Muscle and Nerve* **8**, 355–8.

Lopez, J.R., Alamo, L.A., Jones, D.E., Papp, L., Allen, P.D., Gergely, J. & Sreter, F.A. (1986). $[Ca^{2+}]_i$ in muscles of malignant hyperthermia susceptible pigs determined in vivo with Ca^{2+} selective microelectrodes. *Muscle and Nerve* **9**, 85–6.

Lopez, J.R., Allen, P.D., Alamo, L.A., Jones, D.E. & Sreter, F.A. (1988). Myoplasmic free $[Ca^{2+}]$ during a malignant hyperthermia episode in swine. *Muscle and Nerve* **11**, 82–8.

McCarthy, T.V., Healy, J.M.S., Heffron, J.J.A., Lehane, M., Deufel,

T., Lehmannhorn, F., Farrall, M. & Johnson, K. (1990). Localization of the malignant hyperthermia susceptibility locus to human chromosome 19Q12–13.2. *Nature* **343**, 562–4.

McGrath, C.J. (1986). Malignant hyperthermia. *Seminars in Veterinary Medical Surgery (Small Animal)* **1**, 238–44.

MacLennan, D.H., Duff, C., Zorzato, F., Fujii, J., Phillips, M., Korneluk, R.G., Frodis, W., Britt, B.A. & Worton, R.G. (1990). Ryanodine receptor gene is a candidate for predisposition to malignant hyperthermia. *Nature* **343**, 559–61.

Marjanen, L.A. & Denborough, M.A. (1982). Adenylate kinase and malignant hyperpyrexia. *British Journal of Anaesthesia* **54**, 949–52.

Marjanen, L.A., Shaw, D.C. & Denborough, M.A. (1983). Comparison of adenylate kinase from normal and malignant hyperpyrexic porcine muscle. *Biochemical Medicine* **29**, 164–70.

Mickelson, J.R., Gallant, E.M., Litterer, L.A., Johnson, K.M., Rempel, W.E. & Louis, C.F. (1988). Abnormal sarcoplasmic reticulum ryanodine receptor in malignant hyperthermia. *Journal of Biological Chemistry* **263**, 9310–15.

Mickelson, J.R., Ross, J.A., Hyslop, R.J., Gallant, E.M. & Louis, C.F. (1987*a*). Skeletal muscle sarcolemma in malignant hyperthermia: evidence for a defect in calcium regulation. *Biochimica et Biophysica Acta* **897**, 364–76.

Mickelson, J.R., Thattee, H.S., Beaudry, T.M., Gallant, E.M. & Louis, C.F. (1987*b*). Increased skeletal muscle acetylcholinesterase activity in porcine malignant hyperthermia. *Muscle and Nerve* **10**, 723–7.

Mitchell, G. & Heffron, J.J.A. (1982). Porcine stress syndromes. *Advances in Food Research* **28**, 167–230.

Nelson, T.E. (1988). SR function in malignant hyperthermia. *Cell Calcium* **9**, 257–65.

Nicholls, D.G. (1986). Intracellular calcium homeostasis. *British Medical Bulletin* **42**, 353–8.

Niebroj-Dobosz, I., Kwiatkowski, H. & Mayner-Zawadzka, E. (1984). Experimental porcine malignant hyperthermia: Macromolecular characterisation of muscle plasma membranes. *Medical Biology* **62**, 250–4.

O'Brien, P.J. (1987). Etiopathogenic defect of malignant hyperthermia: hypersensitive calcium release channel of skeletal muscle sarcoplasmic reticulum. *Veterinary Research Communications* **11**, 527–59.

O'Brien, P.J., Forsyth, D.W., Olexson, H.S., Thatte, K.S. & Addis, P.B. (1984). Canine malignant hyperthermia susceptibility: erythrocyte defects, osmotic fragility, glucose-6-phosphate dehydrogenase deficiency and abnormal Ca^{2+} homeostasis. *Canadian Journal of Comparative Medicine* **48**, 381–9.

O'Brien, P.J., Pook, H.A., Klip, A., Britt, B.A., Kalow, B.I., McLaughlin, R.N., Scott, E. & Elliot, M.E. (1990). Canine stress syndrome/ malignant hyperthermia susceptibility: calcium-homeostasis defect in muscle and lymphocytes. *Research in Veterinary Science* **48**, 124–8.

Ohnishi, S.T., Katagi, H., Ohnishi, T. & Brownell, A.K.W. (1988). Detection of malignant hyperthermia susceptibility using a spin label technique on red blood cells. *British Journal of Anaesthesia* **61**, 565–8.

Ohnishi, S.T. & Ohnishi, T. (1988). Halothane induced disorder of red cell membranes of subjects susceptible to malignant hyperthermia. *Cell Biochemistry and Function* **6**, 257–61.

Ohnishi, S.T., Waring, A.J., Fong, S.-R. G., Horivchi, K., Flick, J.L., Sandonaga, K.K. & Ohnishi, T. (1986). Abnormal membrane properties of the sarcoplasmic reticulum of pigs susceptible to malignant hyperthermia: modes of action of halothane, caffeine, dantroline and two other drugs. *Archives of Biochemistry and Biophysics* **247**, 294–301.

Plummer, J.L., Beckwith, A.L.J., Bastin, F.N., Adams, J.F., Cousins, M.J. & Hall, P. (1982). Free radical formation and hepatotoxicity due to anaesthesia with halothane. *Anesthesiology* **57**, 160–6.

Poyer, J., McCay, P.B., Waddle, C.C. & Downs, P.E. (1981). In vivo spin-trapping of radicals formed during halothane metabolism. *Biochemical Pharmacology* **30**, 1517–19.

Richter, C. & Frei, B. (1988). Ca^{2+} release from mitochondria induced by prooxidants. *Free Radical Biology and Medicine* **4**, 365–75.

Rock, E. & Kozak-Reiss, G. (1987). Effect of halothane on the Ca^{2+}-transport system of surface membranes isolated from normal and malignant hyperthermia pig skeletal muscle. *Archives of Biochemistry and Biophysics* **256**, 703–7.

Rosenberg, H. (1988). Clinical presentation of malignant hyperthermia. *British Journal of Anaesthesia* **60**, 268–73.

Rotruck, J.T., Pope, A.L., Ganther, H.E. & Hoekstra, W.G. (1972). Prevention of damage to rat erythrocytes by selenium. *Journal of Nutrition* **102**, 689–96.

Schanus, E.G., Lovrien, R.E. & Taylor, C.A. (1982). Malignant hyperthermia (MH) in humans: deficiencies in the protective systems for oxidative damage. *Progress in Clinical and Biological Research* **97**, 95–111.

Schanus, E.G., Schendel, F., Lovrien, R.E., Rempel, W.E. & McGrath, C. (1981). Malignant hyperthermia (MH): Porcine erythrocyte damage from oxidation and glutathione peroxidase deficiency. In *The Red Cell: Fifth Ann Arbor Conference*, ed. G.J. Brewer, pp. 323–39. New York: Liss.

Schmitt, J., Schmidt, K. & Ritter, H. (1974). Hereditary malignant hyperpyrexia associated with muscle adenylate kinase deficiency. *Humangenetik* **24**, 253–6.

Sessler, D.I. (1986). Malignant hyperthermia. *Journal of Pediatrics* **109**, 9–14.

Sim, A.T.R., White, M.D. & Denborough, M.A. (1987). Effects of adenyl cyclase activators on porcine skeletal muscle in malignant hyperpyrexia. *British Journal of Anaesthesia* **59**, 1557–62.

Stanec, A. & Stefano, G. (1984). Cyclic AMP in normal and malignant hyperpyrexia susceptible individuals following exercise. *British Journal of Anaesthesia* **56**, 1243–6.

Ursini, F., Maiorini, M. & Gregolin, C. (1985). The selenoenzyme phospholipid hydroperoxide glutathione peroxidase. *Biochimica et Biophysica Acta* **839**, 62–70.

Webb, A.J. (1980). The halothane test: A practical method for eliminating the porcine stress syndrome. *Veterinary Record* **106**, 410–12.

Willner, J.H., Cerri, C. & Wood, D.S. (1981). High skeletal muscle adenyl cyclase in malignant hyperthermia. *Journal of Clinical Investigation* **68**, 1119–21.

Younker, D., DeVore, M. & Hartlage, P. (1984). Malignant hyperthermia and glucose-6-phosphate deficiency. *Anesthesiology* **60**, 601–3.

M.J. JACKSON, A. McARDLE
and R.H.T. EDWARDS

Free radicals, calcium and damage in dystrophic and normal skeletal muscle

Introduction

Skeletal muscles are subjected to considerable physical stresses during normal contractile activity, and during excessive or unaccustomed exercise may become seriously damaged such that normal contractile function is impaired. This is evidenced by morphological and ultrastructural changes in muscle together with leakage of large intracellular components (such as certain cytosolic enzymes) into the extracellular fluid. Analogous changes appear to occur in various disease states such as the muscular dystrophies, malignant hyperthermia and various inflammatory myopathies.

In man, muscle damage can be conveniently monitored by measurement of the activity of various muscle-derived enzymes in the blood. The most commonly used of these are creatine kinase or aldolase with creatine kinase determination being particularly useful because of its high sensitivity and because analysis of the isoform pattern of the MM type allows an examination of the elapsed time since the occurrence of an episode of muscle damage leading to enzyme efflux (Page *et al.*, 1989). Further evidence for, or confirmation of, the occurrence of damage can readily be obtained by percutaneous biopsy of the affected muscle under local anaesthetic (Edwards, MacLennan & Jackson, 1989) followed by histological or electron microscopic examination of the tissue. Studies designed to elucidate the mechanisms by which skeletal muscle damage occurs following exercise or in disease states are rare in comparison to studies of the heart or other organs, but a number have been undertaken using different systems. Many of the human studies have (of necessity) been non-invasive and descriptive from which little information concerning basic mechanisms can be obtained, but many further data have been obtained from experiments with animal models *in vivo* and from *in vitro* studies of isolated skeletal muscle tissue.

Evidence for an involvement of calcium in the mechanisms of damage to normal skeletal muscle

Changes in the intracellular calcium content have been implicated in the mechanisms by which damage occurs to several tissues, including skeletal muscle. In cardiac tissue, damage due to hypoxia or reoxygenation has been shown to be associated with an increase in tissue calcium content (Nayler, Poole-Wilson & Williams, 1979), while in hepatocytes loss of cell viability following incubation with various toxins has been reported to be dramatically reduced when the external calcium is removed from the incubation fluid (Schanne *et al.*, 1979); this finding has subsequently been the subject of much controversy (e.g. Smith, Thor & Orrenius, 1981; Farris, Pascoe & Read, 1985).

High external calcium concentrations (3–10 mmols per litre) have been shown to increase CK release from resting animal (Soybell, Morgan & Cohen, 1978) and human (Anand & Emery, 1980) skeletal muscle. Treatment of skeletal muscle preparations with the calcium ionophore A23187 has further demonstrated the potential of increased calcium levels to induce damage (Duncan, Smith & Greenaway, 1979). Slow calcium channel blocking agents (i.e. calcium antagonists) have been found to reduce the CK release from human skeletal muscle *in vitro* (Anand & Emery, 1982) and other workers have shown that alternative manipulations designed to reduce calcium accumulation in skeletal muscle (i.e. parathyroidectomy) prevents the pathological changes to skeletal muscle in hamsters with an inherited form of muscle degeneration (Palmieri *et al.*, 1981). In another pathological condition of muscle (selenium deficiency myopathy) ^{45}Ca accumulation by muscles has been reported to precede biochemical, histological, or clinical evidence of myopathy (Godwin, Edwardly & Fuss, 1975).

We have used an *in vitro* system to examine the role of external calcium in the release of enzymes from skeletal muscle (Jones, Jackson & Edwards, 1983). It was found that release of enzymes following different stresses (e.g. excessive contractile activity, treatment with low-dose detergents, or treatment with mitochondrial inhibitors) could be prevented by removal of the external calcium during the damaging procedure (Jones *et al.*, 1984; Jackson, Jones & Edwards, 1984). It was also found that this manipulation was equally effective in protection of muscle against the histological and electron microscopic changes induced by excess contractile activity (Jones *et al.*, 1984). Other experiments with this system have also demonstrated a dramatic increase in total muscle calcium during either excess contractile activity, leading to enzyme efflux, or treatment with mitochondrial inhibitors (e.g. dinitrophenol) (Claremont, Jackson & Jones, 1984).

These results suggest that damage to skeletal muscle is accompanied by an influx of extracellular calcium down the large extracellular to intracellular concentration gradient for this element. This increased intracellular calcium content then mediates further pathological changes. Furthermore we have recently demonstrated that an accumulation of intracellular calcium can cause damage to skeletal muscle without a concomitant failure of muscle energy supply (West-Jordan *et al.*, 1990); however, despite the fact that others have claimed an effect of calcium antagonists in skeletal muscle (Anand & Emery, 1982), we have been unable to demonstrate any protective effect of these agents in isolated preparations (Jones *et al.*, 1984).

Considerable speculation has surrounded the possible mechanisms by which increased calcium may mediate pathological processes in cells. The hypotheses which have been proposed regarding skeletal muscle include accumulation of calcium by mitochondria leading to loss of oxidative energy production (Wrogemann & Pena, 1976), activation of calcium dependent proteases (Ebashi & Sugita, 1979), activation of lysosomal proteases by the stimulation of prostaglandin production (Rodemann, Waxman & Goldberg, 1981) or direct release of lysosomal enzymes (Duncan, 1978).

Inhibitor studies which we have performed suggest that calcium influx may be a key step in the damage resulting in an activation of phospholipase A (Jackson *et al.*, 1984). Activation of this enzyme will result in the breakdown of membrane phospholipids leading to production of lysophospholipids and free fatty acids. Accumulation of lysophospholipids will lead to a breakdown of membrane lipid organisation (Weglicki, 1980) and the free fatty acids released will act as detergents, causing membrane damage (Katz, 1982). In addition, among the free fatty acids produced will be arachidonic acid. This is the precursor of the prostaglandin series of compounds. Prostaglandins have been reported to be involved in the control of muscle protein homeostasis (Rodemann *et al.*, 1981; Baracos *et al.*, 1983) and prostaglandins E_2 and $F_{2\alpha}$ have been shown to be released by skeletal muscle following an accumulation of intracellular calcium (Jackson, Wagenmakers & Edwards, 1987), but these substances do not appear to be directly involved in the mechanism by which calcium damages muscle cells.

Role of free radicals in damage to normal muscle cells

Several workers have implicated an increase in free-radical-mediated reactions in the damage to muscle which accompanies exercise. Brady and co-workers (1979) studied the response of rats to exhaustive swimming exercise and found that both liver and muscle tissue contained increased amounts of malondialdehyde (a product of free-radical-mediated lipid peroxidation) following exercise and suggested that an increased amount of free-radical intermediates may be produced during exercise. Tappel's group has studied pentane excretion in the breath of normal subjects (Dillard *et al.*, 1978) and rats (Gee & Tappel, 1981) during exercise (pentane is a product of the free-radical-mediated peroxidation of certain fatty acids) and has found a large increase, suggesting increased lipid peroxidation at some site in the body during exercise. Dillard *et al.* (1978) have also demonstrated an apparent protective effect of supplemental vitamin E against this process. The major role of vitamin E in the body appears to be to act as a lipid-soluble antioxidant preventing free-radical-mediated peroxidation of membrane components. Packer and co-workers have been specifically examining the effects of exercise in vitamin E deficient animals (Quintanilha & Packer, 1983), and they claim that these animals have a considerably lower tolerance to exercise since they are less able to withstand the oxidative stress which occurs with increased mitochondrial energy metabolism. These workers have also demonstrated increased levels of lipid peroxidation products (malonaldehyde) in animal tissues following exercise (Davies *et al.*, 1982).

It has been pointed out that damage to tissues can be the cause of lipid peroxidation as well as the consequence of it (Halliwell & Gutteridge, 1984), and the studies examining lipid peroxidation products following exercise may well fall into this category. An alternative approach is to try and examine free-radical intermediates directly in tissues using physical methods. Davies *et al.* (1982) have shown that strenuous exercise leads to an increased electron spin resonance signal from muscle and liver tissue. We have also examined electron spin resonance signals from skeletal muscle during experimental skeletal muscle damage induced by excessive contractile activity *in vivo* (Jackson, Edwards & Symons, 1985). In these studies, a 70 ± 20% increase in the stable electron spin resonance signal was shown to be associated with an increase in the plasma CK activity of the rat following exercise. Unfortunately, in this situation it is still not possible to say which of the increased free-radical concentration or the damage to the muscle is primary or whether the two findings are merely coincidental.

We have also used the *in vitro* skeletal muscle damage system (Jones *et al.*, 1983) to examine this area. It has been demonstrated that the vitamin E content of the muscle influences the amount of lactate dehydrogenase enzyme released from muscles following an equivalent amount of contractile activity both *in vitro* and *in vivo* (Jackson, Jones & Edwards, 1983). Recent data have cast some doubt on the interpretation of these results since Phoenix, Edwards & Jackson (1989; 1990) have demonstrated that α-tocopherol (vitamin E) is protective against calcium-induced damage to skeletal muscle and furthermore that the protective effects are mimicked by α-tocopherol acetate, phytol and isophytol which have no antioxidant properties in this system.

Muscle damage in the muscular dystrophies

Duchenne and Becker muscular dystrophy are chronic, degenerative muscle-wasting disorders for which there is currently no known therapy. Recent studies have identified and characterised the defective gene responsible for these disorders (Monaco *et al.*, 1985) and the protein for which this gene normally codes has been identified and named 'dystrophin' (Hoffmann, Brown & Kunkel, 1987). Studies using antibodies raised against part of the protein suggest that it is deficient in muscle from patients with Duchenne muscular dystrophy (DMD) and present in reduced quantities or of abnormal size in many patients with Becker muscular dystrophy (Hoffmann *et al.*, 1987). The implication of these findings is that a lack of this protein leads to muscle degeneration in this disorder although the mechanism by which this occurs has not been elucidated. Dystrophin appears to be localised under the plasma membrane (Watkins *et al.*, 1988) and has been proposed to exert a stabilising effect on the muscle membrane during normal contraction, a protective effect which is missing in dystrophin-deficient dystrophic muscle (Karpati & Carpenter, 1988). Alternative theories put forward for the function of dystrophin are that it is involved in normal calcium regulation in muscle (Hoffmann *et al.*, 1987), in stabilisation of the stretch-activated ion channels of the muscle plasma membrane (Duncan, 1989) or in stabilising certain glycoproteins in the muscle membrane (Ervasti *et al.*, 1990).

The *mdx* mouse model of muscular dystrophy has also been shown to lack dystrophin although it displays only minor evidence of muscle damage and shows no apparent 'clinical' dysfunction of the musculature (Anderson, Ovalle & Bressler, 1987). However, the *mdx* mouse does display a consistent elevation of circulating muscle-derived enzymes, such as creatine kinase (CK) and an abnormal elevation of muscle protein degradation *in vitro* (Turner *et al.*, 1988) and both synthesis and degradation rates *in vivo* (MacLennan & Edwards, 1990). *Mdx* muscle also contains a higher content of intracellular calcium than control muscles (Turner *et al.*, 1988) and this increased calcium appears to mediate increased tyrosine release from the *mdx* muscle *in vitro*, thus suggesting a crucial mediating role for intracellular calcium in the degeneration in dystrophic animal muscle.

We have examined the possible role of extracellular calcium in mediating the damage to dystrophic skeletal muscle by *in vitro* studies of strips of biceps muscle from patients with Duchenne muscular dystrophy and control subjects (Jackson *et al.*, 1991). These data demonstrate an elevated release of creatine kinase from DMD muscle *in vitro* accompanied by a grossly elevated release of prostaglandin E_2. Both of these are reduced during the time of incubation with the creatine kinase release becoming equal to normal after 90 min of incubation. Immediate immersion of the muscle strips in calcium-free medium also reduced the difference between control and dystrophic samples supporting a mediating role for calcium in the damage to dystrophic muscle.

In recent experiments we have attempted to undertake similar experiments to examine the mechanisms by which loss of intracellular enzymes occurs from muscle of the *mdx* mouse *in vitro* (McArdle, Edwards & Jackson, 1991). Surprisingly these studies have failed to demonstrate an elevated efflux *in vitro* despite an apparent elevated efflux *in vivo* and furthermore demonstrate a reduced susceptibility of *mdx* muscle to damaging contractile activity. However, prostaglandin E_2 release by *mdx* muscle in response to damaging contractile activity was elevated compared to control tissue following repetitive contractile activity either with or without simultaneous stretching.

These latter findings are somewhat difficult to interpret in the light of previous results. It is clear from both studies with *mdx* mouse muscle and strips of Duchenne muscle that dystrophic muscle releases increased amounts of prostaglandin E_2. Simultaneous analyses of leukotriene B_4 in the incubation media suggests that this prostaglandin is derived from muscle tissue rather than non-muscle (infiltrating) cells. This prostaglandin is likely to be formed from arachidonic acid released by phospholipase action on membrane phospholipids and may reflect an increased calcium content of the dystrophic muscle although the stimulatory effects of the calcium ionophore are suggestive of some overactivity of phospholipase catalysed phospholipid degeneration in dystrophic muscle in response to an equivalent increase in calcium content (Jackson *et al.*, 1991; McArdle *et al.*, 1991). Much more difficult to interpret is the lack of elevated creatine kinase release from the mdx muscle *in vitro* compared to control (McArdle *et al.*, 1991) in comparison to the studies with strips of human DMD biceps (Jackson *et al.*, 1991).

However, since the elevated CK release from the human DMD muscle is also reduced to normal with time of incubation *in vitro* it is possible that there is some factor *in vivo* which is necessary for the continued release of intracellular components from dystrophic muscle and which is lost when muscles are studied *in vitro*.

Acknowledgements

The authors would like to thank the Muscular Dystrophy Group of Great Britain and Northern Ireland for continuing financial support and the many collaborators and colleagues who have contributed to the work described.

References

Anand, R. & Emery, A.E.H. (1980). Calcium stimulated enzyme efflux from human skeletal muscle. *Research Communications in Chemical Pathology and Pharmacology* **28**, 541–50.

Anand, R. & Emery, A.E.H. (1982). Verapamil and calcium-stimulated enzyme efflux from skeletal muscle. *Clinical Chemistry* **28**, 1482–4.

Anderson, J.E., Ovalle, W.K. & Bressler, B.H. (1987). Electron microscopic and autoradiographic characterisation of hindlimb muscle regeneration in the mdx mouse. *Anatomy Records* **219**, 243–57.

Baracos, V., Rodemann, P., Dinarello, C.A. & Goldberg, A.L. (1983). Stimulation of muscle protein degradation and prostaglandin E_2 release by leucocyte pyrogen (Interleukin-1). *New England Journal of Medicine* **308**, 553–8.

Brady, P.S., Brady, W. & Ullreg, D.E. (1979). Selenium, vitamin E and the response to swimming stress in the rat. *Journal of Nutrition* **109**, 1103–9.

Claremont, D., Jackson, M.J. & Jones, D.A. (1984). Accumulation of calcium in experimentally damaged mouse muscles. *Journal of Physiology* **353**, 57P.

Davies, K.J.A., Quintanilha, A.T., Brooks, G.A. & Packer, L. (1982). Free radicals and tissue damage produced by exercise. *Biochemistry and Biophysics Research Communications* **107**, 1198–1205.

Dillard, C.J., Litov, R.E., Savin, W.M., Dunelin, E.E. & Tappel, A.L. (1978). Effects of exercise, vitamin E and ozone on pulmonary function and lipid peroxidation. *Journal of Applied Physiology* **45**, 927–32.

Duncan, C.J. (1978). Role of intracellular calcium in promoting muscle damage: a strategy for controlling the dystrophic condition. *Experientia* **34**, 1531–5.

Duncan, C.J. (1989). Dystrophin and the integrity of the sarcolemma in Duchenne muscular dystrophy. *Experientia* **45**, 175–7.

Duncan, C.J., Smith, J.L. & Greenaway, H.C. (1979). Failure to protect frog skeletal muscle from ionophore-induced damage by the use of the protease inhibitor leupeptin. *Comparative Biochemistry and Physiology* **63**C, 205.

Ebashi, S. & Sugita, H. (1979). The role of calcium in physiological and pathological processes of skeletal muscle. In *Current Topics in Nerve and Muscle Research*, ed. A.J. Aguayo & G. Karpati, pp. 73–86. Exerpta Medica.

Edwards, R.H.T., MacLennan, P. & Jackson, M.J. (1989). Myochemistry in myopathy. In *Advances in Myochemistry: 2*, pp. 135–49. John Libbey Eurotext.

Ervasti, J.M., Chlendieck, K., Kahl, S.D., Graver, M.G. & Campbell, K.P. (1990). Deficiency of a glycoprotein component of the dystrophin complex in dystrophic muscle. *Nature* **345**, 315–19.

Farris, M.W., Pascoe, G.A. & Reed, D.J. (1985). Vitamin E reversal of the effect of extracellular calcium on chemically induced toxicity in hepatocytes. *Science* **277**, 751–4.

Gee, D.L. & Tappel, A.L. (1981). The effect of exhaustive exercise on expired pentane as a measure of 'in vivo' lipid peroxidation in the rat. *Life Sciences* **28**, 2425–9.

Godwin, K.O., Edwardly, J. & Fuss, C.N. (1975). Retention of ^{45}Ca in rats and lambs associated with the onset of nutritional muscular dystrophy. *Australian Journal of Biological Sciences* **28**, 457–60.

Halliwell, B. & Gutteridge, J.M.C. (1984). Lipid peroxidation oxygen radicals, cell damage and antioxidant therapy. *Lancet* **ii**, 1396–7.

Hoffmann, E.P., Brown, R.H. & Kunkel, L.M. (1987). Dystrophin: the protein product of the Duchenne muscular dystrophy locus. *Cell* **51**, 919–28.

Jackson, M.J., Brooke, M.H., Kaiser, K. & Edwards, R.H.T. (1991). Creatine kinase and prostaglandin E_2 release from isolated Duchenne muscle. *Neurology* **41**, 101–4.

Jackson, M.J., Edwards, R.H.T. & Symons, M.C.R. (1985). Electron spin resonance studies of intact mammalian skeletal muscle. *Biochimica et Biophysica Acta* **847**, 185–90.

Jackson, M.J., Jones, D.A. & Edwards, R.H.T. (1983). Vitamin E and skeletal muscle. In *Biology of Vitamin E Ciba Foundation Symposium 101*, pp. 224–39. London: Pitman.

Jackson, M.J., Jones, D.A. & Edwards, R.H.T. (1984). Experimental skeletal muscle damage: the nature of the calcium-activated degenerative processes. *European Journal of Clinical Investigation* **14**, 369–74.

Jackson, M.J., Wagenmakers, A.J.M. & Edwards, R.H.T. (1987). The effect of inhibitors of arachidonic acid metabolism on efflux of intracellular enzymes from skeletal muscle following experimental damage. *Biochemical Journal* **241**, 403–7.

Jones, D.A., Jackson, M.J. & Edwards, R.H.T. (1983). The release of intracellular enzyme from an isolated mammalian skeletal muscle preparation. *Clinical Science* **65**, 193–201.

Jones, D.A., Jackson, M.J., McPhail, G. & Edwards, R.H.T. (1984). Experimental muscle damage: The importance of external calcium. *Clinical Science* **66**, 317–22.

Karpati, G. & Carpenter, S.C. (1988). The deficiency of a sarcolemmal cytoskeletal protein (dystrophin) leads to necrosis of skeletal muscle fibres in Duchenne–Becker dystrophy. In *Neuromuscular Junction*, ed. L.S. Sellin, R. Libelius & S. Thesleff, pp. 429–36. Elsevier.

Katz, A.M. (1982). Membrane-derived lipids and the pathogenesis of ischaemic myocardial damage. *Journal of Molecular and Cell Cardiology* **14**, 627–32.

McArdle, A., Edwards, R.H.T. & Jackson, M.J. (1991). Effects of contractile activity on indicators of muscle damage in the dystrophin-deficient mdx mouse. *Clinical Science* (in press).

MacLennan, P. & Edwards, R.H.T. (1990). Protein turnover is elevated in muscle of mdx mice 'in vivo'. *Biochemical Journal* **268**, 795–7.

Monaco, A.P., Berteson, C.J., Middlesworth, W., Colletti, C.-A., Aldridge, A., Fischbeck, K.H., Bartlett, R., Pericak-Vance, M.A., Roses, A. & Kurkel, L.M. (1985). Detection of deletions spanning the Duchenne muscular dystrophy locus using a tightly linked DNA segment. *Nature* **316**, 842–5.

Nayler, W.G., Poole-Wilson, P.A. & Williams, A. (1979). Hypoxia and calcium. *Journal of Molecular and Cell Cardiology* **11**, 683–706.

Page, S., Jackson, M.J., Coakley, J. & Edwards, R.H.T. (1989). Creatine kinase MM isoforms in the study of skeletal muscle damage. *European Journal of Clinical Investigation* **19**, 185–91.

Palmieri, G.M.A., Nutting, D.F., Bhattacharya, S.K., Bartorini, T.E. & Williams, J.C. (1981). Parathyroid ablation in dystrophic hamsters. *Journal of Clinical Investigation* **68**, 646–54.

Phoenix, J., Edwards, R.H.T. & Jackson, M.J. (1989). Inhibition of calcium-induced cytosolic enzyme efflux from skeletal muscle by vitamin E and related compounds. *Biochemical Journal* **287**, 207–13.

Phoenix, J., Edwards, R.H.T. & Jackson, M.J. (1990). Effects of calcium ionophore on vitamin E deficient muscle. *British Journal of Nutrition* **64**, 245–56.

Quintanilha, A.T. & Packer, L. (1983). Vitamin E, physical exercise and tissue oxidative damage. In *Biology of Vitamin E Ciba Foundation Symposium 101*, ed. R. Porter & J. Whelan, pp. 56–69. London: Pitman.

Rodemann, H.P., Waxman, L. & Goldberg, A.L. (1981). The stimulation of protein degradation in muscle by Ca^{2+} is mediated by prostaglandin E_2 and does not require the calcium-activated protease. *Journal of Biological Chemistry* **257**, 8716–23.

Schanne, F.X., Kane, A.B., Young, A.B. & Forber, J.L. (1979). Calcium dependence of toxic cell death: a final common pathway. *Science* **206**, 700–1.

Smith, M.J., Thor, H. & Orrenius, S. (1981). Toxic injury to isolated hepatocytes is not dependent on extracellular calcium. *Science* **213**, 1257–9.

Soybell, D., Morgan, J. & Cohen, L. (1978). Calcium augmentation of enzyme leakage site of action. *Research Communications in Chemical Pathology and Pharmacology* **20**, 317–29.

Turner, P.R., Westwood, T., Regan, C.M. & Steinhardt, R.A. (1988). Increased protein degradation results from elevated free calcium levels found in muscle from mdx mice. *Nature* **335**, 735–8.

Watkins, S.C., Hoffman, E.P., Slayter, H.S. & Kunkel, L.M. (1988). Immunoelectron microscopic localisation of dystrophin in myofibres. *Nature* **333**, 863–6.

Weglicki, W.B. (1980). Degradation of phospholipids in myocardial membranes. In *Degratative Processes in Heart and Skeletal Muscle*, ed. K. Wilderthal, pp. 377–88. Elsevier.

West-Jordan, J., Martin, P., Abraham, R.J., Edwards, R.H.T. & Jackson, M.J. (1990). Energy dependence of cytosolic enzyme efflux from skeletal muscle. *Clinica et Chimica Acta* **189**, 163–72.

Wrogemann, K. & Pena, S.J.G. (1976). Mitochondrial overload: a general mechanism for cell necrosis in muscle disease. *Lancet* **ii**, 672–4.

C.J. DUNCAN and NASRIN SHAMSADEEN

Ultrastructural changes in mitochondria during rapid damage triggered by calcium

Ultrastructural changes in muscle mitochondria during cell damage

During rapid cellular damage, the organelles frequently undergo major ultrastructural pathological changes which are shown most dramatically in the mitochondria. In particular, the mitochondria undergo apparent septation and subdivision, this phenomenon being most commonly seen in skeletal and cardiac muscle cells (Duncan, 1988). Are these ultrastructural changes in muscle mitochondria perhaps triggered by changes in the intracellular concentration of Ca^{2+} ($[Ca^{2+}]_i$) or by active oxygen radicals?

Lipid bodies, lipofuscin granules and myelin-like figures (or membrane whorls) are regarded as late-stage products of lysosomal digestion or as lysosome-derived elements in mammalian skeletal muscle; they occur rarely in human healthy muscle but are much more common in diseased muscle and in old age (Mastaglia & Walton, 1982; Dubowitz, 1985; Walton, 1988). Lipid droplets are present in many cells and are considered to lack a limiting membrane (Threadgold, 1976), but lipid bodies with a bounding membrane are clearly evident in the electron-micrographs of skeletal muscles of neonatal kittens (Tomanek, 1976), rats (Nag & Cheng, 1982) and dystrophic hamsters (Caulfield, 1966) as well as in ischaemic canine muscle (Stenger *et al.*, 1962). Ultrastructural studies of mammalian skeletal muscle undergoing rapid cellular damage that is experimentally-induced *in vitro*, with a time-course of minutes (Duncan, 1988), show that lipid bodies develop quickly in association with the muscle mitochondria. The damage of the myofilament apparatus exhibits two specific patterns of response that are clearly different, but may occur in adjacent sarcomeres of the same myofibril, namely (i) severely contracted sarcomeres with blurred Z-lines and (ii) relaxed sarcomeres with dissolution of the actin filaments and with the Z-lines remaining sharply defined; this damage eventually leads to the dissolution of complete sarcomeres (Duncan, 1987).

Concomitantly, the myelin-like figures and lipid bodies develop within the mitochondria and it is suggested here that they are not necessarily the products of lysosomal digestion. The status of the lysosomal system and its role in cellular degradation in skeletal muscle is not clear, but there is evidence that some of its functions are served by the sarcotubular system (Duncan, Greenaway & Smith, 1980). The sarcotubular system swells greatly when muscles are incubated in lysosomotropic agents, such as leucine methylester, and the muscle cells eventually undergo the characteristic cellular degradation, but the inhibitors of lysosomal cathepsins are ineffective in ameliorating this damage (Duncan, 1987; Duncan & Rudge, 1988; Shamsadeen & Duncan, 1989) suggesting that acid hydrolases are not implicated in the initial destruction of the myofilaments. The action of lysosomotropic agents in promoting the formation of myelin-like figures and lipid bodies in skeletal muscle has therefore been examined.

The ultrastructure of the myelin-like figures in damaged muscle bears a close resemblance to the multilamellar bodies found in Type II cells of the alveolar epithelium of the mammalian lung which are responsible for the formation and secretion of surfactant (Kikkawa & Spitzer, 1969; Chevalier & Collet, 1973); a deficiency of surfactant is the primary factor in the development of respiratory distress syndrome in the premature newborn. The elucidation of the sequence of events in the rapid formation of myelin-like bodies in muscle cells may also provide useful clues concerning the mode of secretion of lung surfactant dipalmitoyl-phosphatidylcholine (dipalmitoyl-PC).

Material and methods

The possible sequence of events that culminate in the development of myelin figures and lipid bodies has been studied in incubated mouse diaphragm but comparative studies with other rodent skeletal muscles have been included. The procedures for incubation of muscle *in vivo* and for electron microscopy have been given in detail elsewhere (Shamsadeen & Duncan, 1989). Preparations were exposed to a variety of treatments that are believed to raise $[Ca^{2+}]_i$ (see Duncan, 1988) or to lysosomotropic agents (Shamsadeen & Duncan, 1989).

Suggested sequence of ultrastructural changes in mitochondria

Mitochondrial division

Exposure of the mouse diaphragm to agents or conditions that are believed to raise the concentration of intracellular calcium ($[Ca^{2+}]_i$) produced the characteristic rapid cellular damage. Examples of these treatments are (i) A23187, 40–60 min, (ii) caffeine, 10^{-2}M, 40 min, (iii) dinitrophenol, 10^{-3}M, 40 min, (iv) hypoxia. In addition to the characteristic damage of the myofilament apparatus (Publicover, Duncan & Smith, 1978) the mitochondria swell, change their configurational state, develop internal septa, and apparently subdivide and increase in number (Publicover, Duncan & Smith, 1977; Duncan, 1988), see Figs 1 and 2. Some mitochondria develop characteristic internal mitochondrial bars which have a laminar structure and which develop from the cristae, finally becoming electron dense (Fig. 1). Similar changes in the organisation of the mitochondrial cristae were also found in cardiac myocytes undergoing cellular damage (Fig. 3).

Myelin figures

The modification of the mitochondrial membranes to form the conspicuous myelin figures was found more rarely but Figs 4–12 illustrate a possible sequence, showing a progressive increase in the membrane whorls. These mitochondria were from damaged cells of both cardiac and skeletal muscle and it is evident that myelin-like figures develop within the mitochondria and that they can form rapidly, within 40 min.

Myelin-like figures apparently developed from either the membranes of the cristae (Fig. 7) or from the inner and outer mitochondrial membranes (Figs 4–6), although in the latter case these may be from newly-formed septa which subdivide the swollen mitochondria. Although myelin-like figures were found in muscle cells damaged by a variety of treatments, they were particularly conspicuous following relatively long exposure to the lysosomotropic agents leucine methylester or dodecyl imidazole.

Lipid bodies

Lipid bodies are found much more frequently in damaged muscle cells and are normally packed between the mitochondria (Figs 1 and 13). The series of Figs 14–18 represents a suggested sequence of events in the formation of these membrane-bound organelles, showing that they develop from the internal membrane systems of damaged mitochondria, frequently from myelin-like figures. It is suggested that initially the mitochondria are filled with membrane structures and whorls derived from modified cristae or with myelin-like figures (Figs 9, 10, 12, 14, 15, 18, 19 and 20) and that these break down to produce lipid and phospholipid concentrations within the original mitochondrial membrane (Figs 10 and 11). In some examples the membranes break down in the

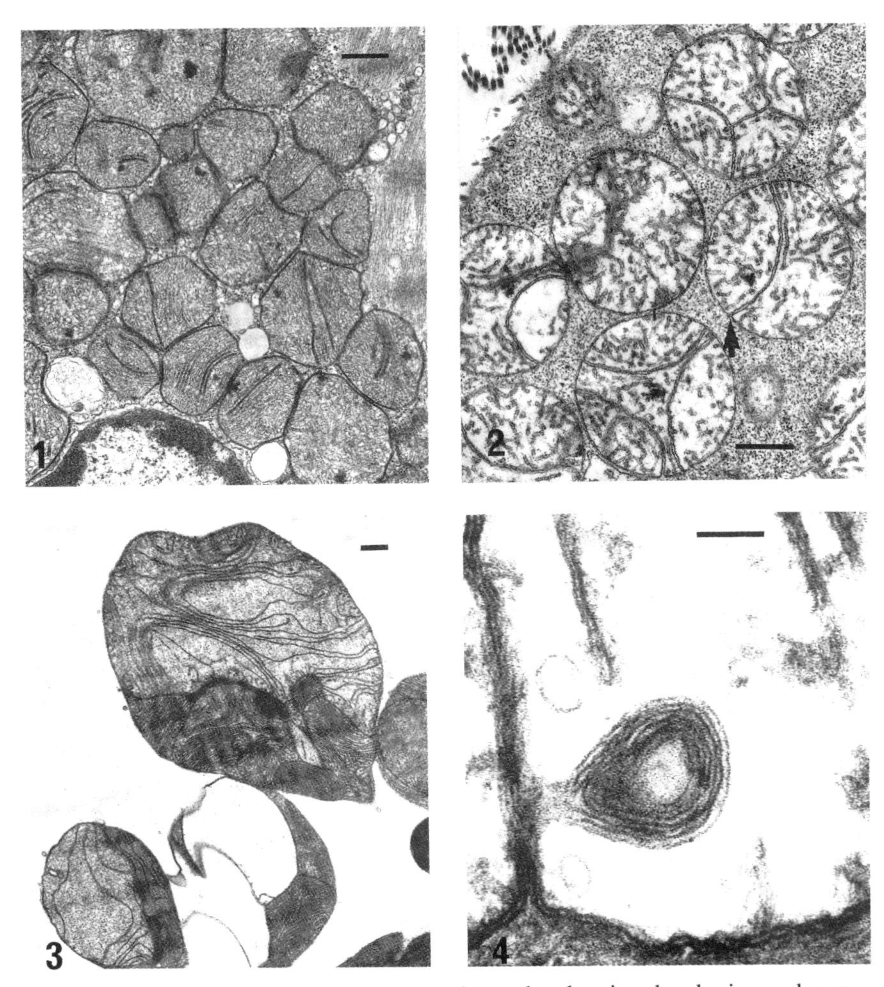

Fig. 1. Mitochondria in damaged muscle, showing developing and complete septa, mitochondrial bars and lipid bodies. Mouse diaphragm, hypoxia, 90 min. Bar = 0.5 μm.

Fig. 2. Mitochondria swollen and subdividing, with septa formed from the inner mitochondrial membrane (arrow). Organisation of cristae severely damaged. Cytosol clearly damaged with loss of myofilament apparatus. Mouse soleus muscle; 10 mM leucine methylester, 3 hr. Bar = 0.5 μm.

Fig. 3. Mitochondria with bizarre modification of cristal membranes forming internal subdivisions and septa which in the upper mitochondria are developing into lamellar figures. Rat cardiomyocytes; 30 μg ml^{-1} dodecyl imidazole. Bar = 0.5 μm.

Fig. 4. Swollen, damaged and subdivided mitochondrion with a myelin figure developing from an internal septum. Mouse diaphragm, hypoxia, 40 min. Bar = 0.1 μm.

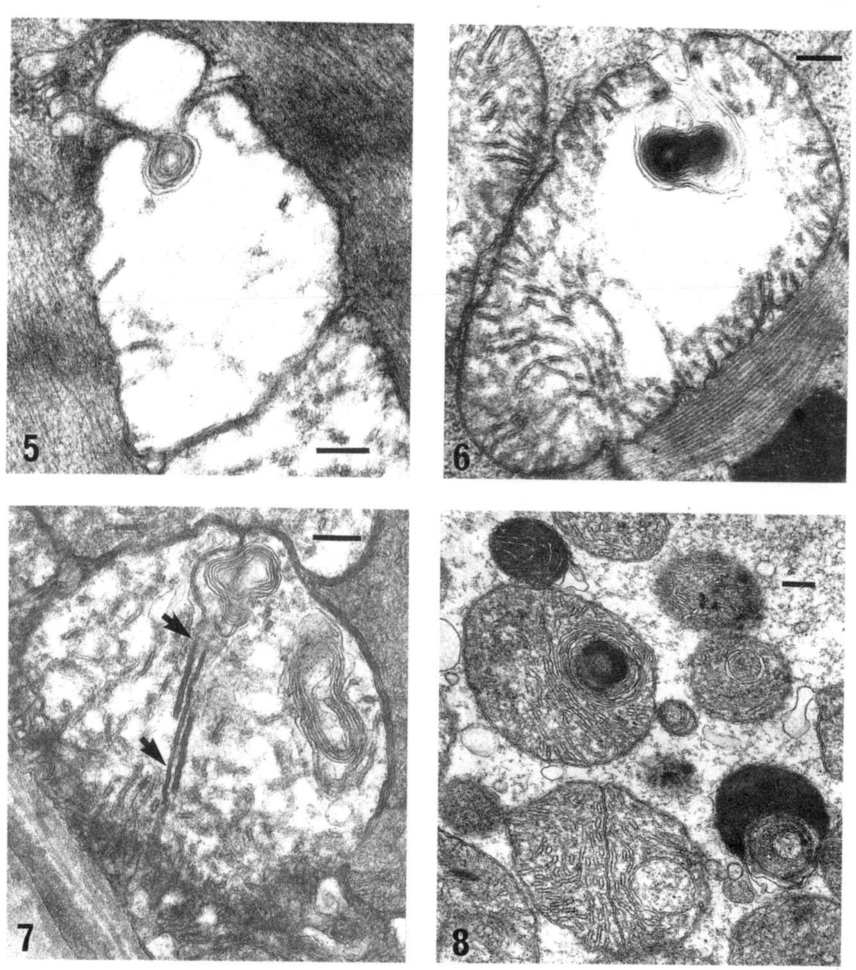

Fig. 5. Myelin figure developing within a swollen and subdivided mitochondrion. Damaged myofilament apparatus. Mouse diaphragm, hypoxia, 40 min. Bar = 0.2 μm.

Fig. 6. Swollen mitochondria with damaged cristae showing a myelin figure developing from the inner and outer mitochondrial and cristal membranes. Mouse diaphragm; 6 mM leucine methylester, 2 hr. Bar = 0.2 μm.

Fig. 7. Damaged mitochondrion showing two myelin figures, probably developed from the cristae which are also forming electron-dense mitochondrial bars (arrows). Mouse diaphragm; 10 mM leucine methylester, 2 hr. Bar = 0.2 μm.

Fig. 8. Damaged mitochondria (one with a complete internal septum) with various internal developments, including myelin figures and lipid bodies. Severely damaged rat cardiomyocytes; 30 min incubation. Bar = 0.2 μm.

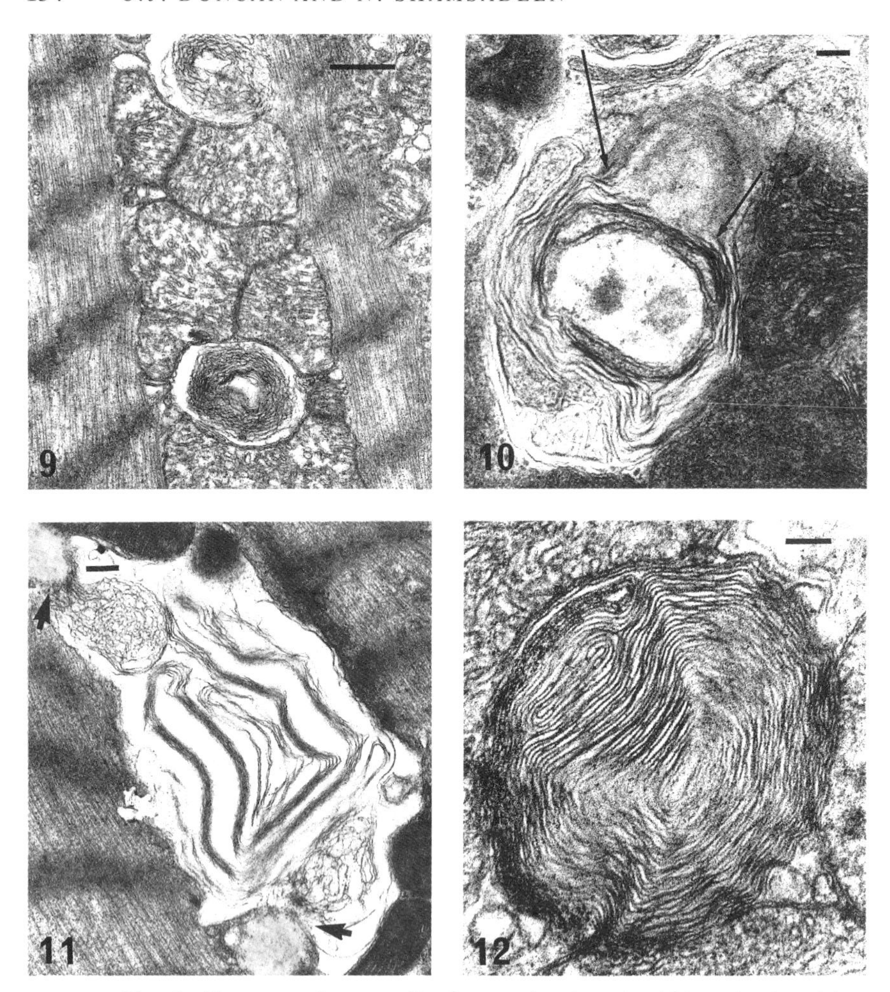

Fig. 9. Two complete myelin figures developed within mitochondria. Damaged myofilament apparatus with blurred Z-lines. Mouse diaphragm; 10 mM leucine methylester, 3 hr. Bar = 0.5 μm.

Fig. 10. Complete myelin figure formed between the mitochondria; the ends of the lamellar membranes are breaking down into the developing lipid body (arrows). Rat cardiomyocytes; 10 mM leucine methylester, 5 hr. Bar = 0.1 μm.

Fig. 11. Myelin figure in which the lamellae spread from damaged cristae. Lipid bodies are developing in association with the damaged membrane (arrows). Mouse diaphragm 50 μg ml^{-1} dodecyl imidazole, 90 min. Bar = 0.2 μm.

Fig. 12. Fully-formed myelin figure with complex membrane whorls between mitochondria. Mouse diaphragm; 60 μg ml^{-1} dodecyl imidazole, 4 hr. Bar = 0.1 μm.

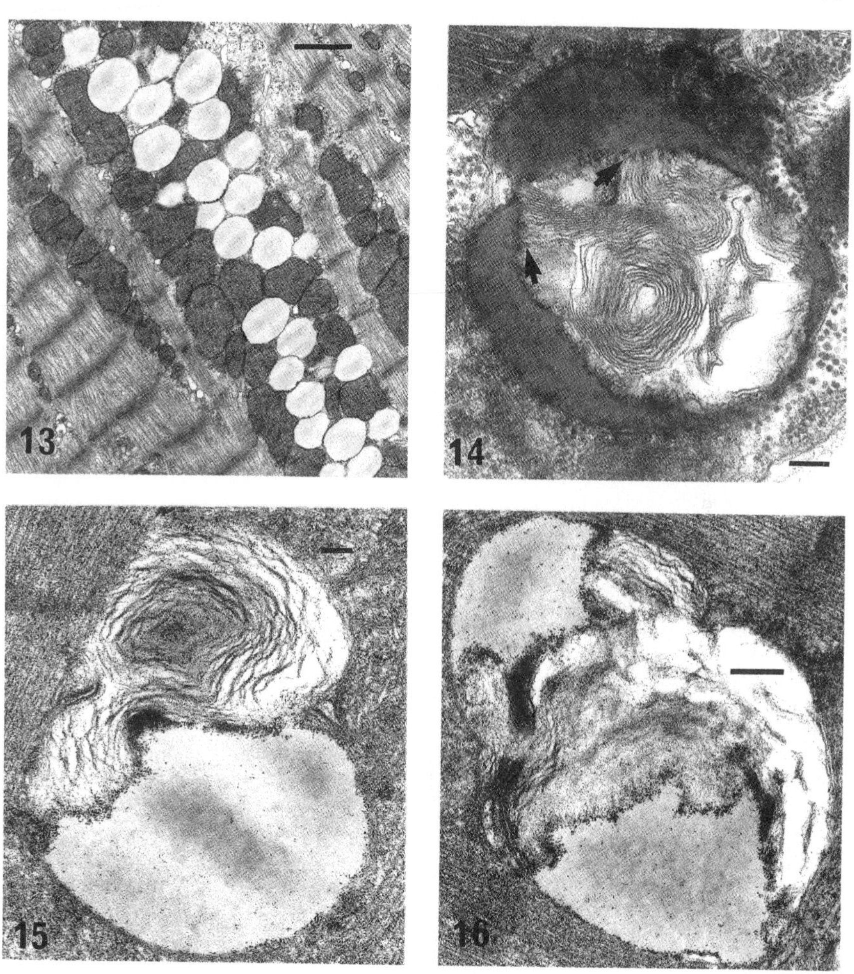

Fig. 13. Multiple, complete lipid bodies packed between mitochondria; damaged myofibrils with blurred Z-lines. Mouse diaphragm; anoxia (N_2 gassing), 40 min. Bar = 1 μm.

Fig. 14. Lipid body forming in two discrete sections at the periphery in close association with myelin figures; the ends of the lamellae are applied to the boundary of the lipid (arrows). Mouse diaphragm; 60 μg ml^{-1} dodecyl imidazole. Bar = 0.1 μm.

Fig. 15. Lipid body forming in association with a lamellated myelin figure. Reaction product for cathepsin B localised at the interface. Mouse diaphragm; 10 mM leucine methylester, 60 min. Bar = 0.1 μm.

Fig. 16. Lipid body forming from the breakdown of internal membranes at two separate sites at the periphery; the interface is blurred and has the reaction product for cathepsin B localised there. Mouse diaphragm; 10 mM leucine methylester, 60 min. Bar = 0.1 μm.

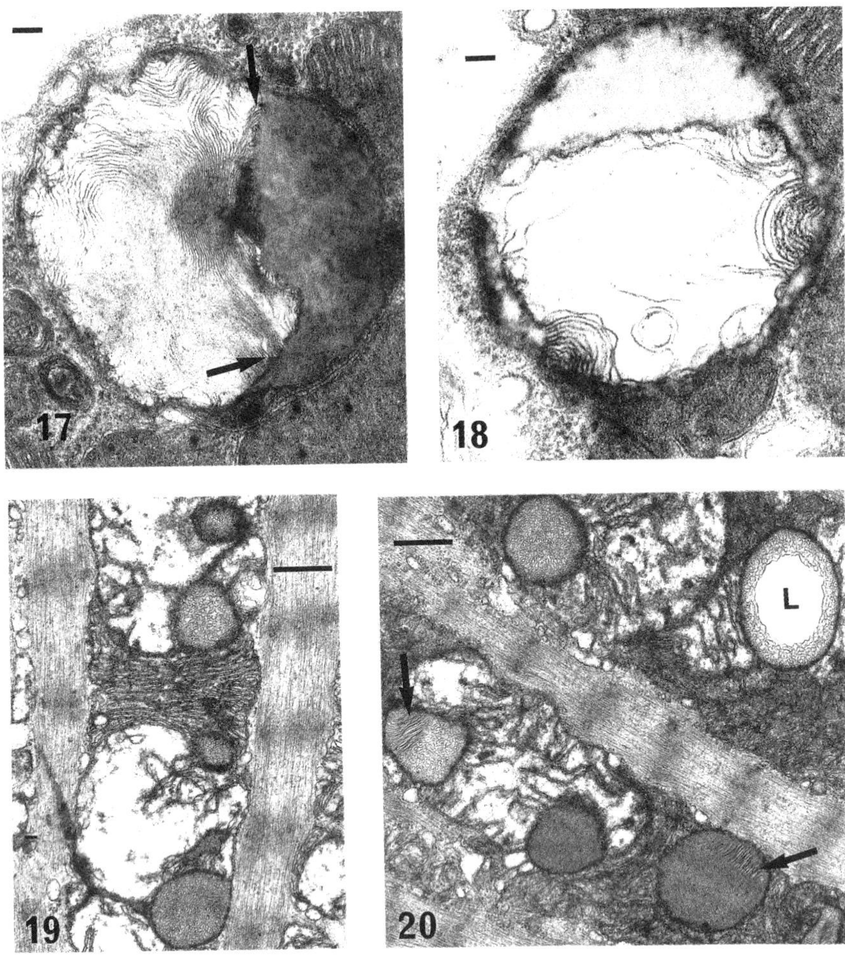

Fig. 17. Lipid body formation with the ends of the lamellae attached to the interface between the two sections. Mouse diaphragm; 60 μg ml^{-1} dodecyl imidazole, 2 hr. Bar = 0.1 μm.

Fig. 18. Lipid body formation at the periphery, with few residual membrane whorls remaining. Mouse diaphragm; 50 μg ml^{-1} dodecyl imidazole, 90 in. Bar = 0.1 μm.

Fig. 19. Swollen and subdivided mitochondria showing sections in which the cristal membranes are modified and packed together in a characteristic way. Mouse diaphragm; 50 μg ml^{-1} dodecyl imidazole, 90 min. Bar = 0.5 μm.

Fig. 20. Mitochondria showing progressive changes in the modifications of the cristae which are still discernible (arrows), then becoming tightly packed and finally showing the start of lipid body formation (L) by the breakdown of the central lamellae. Mouse diaphragm; 50 μg ml^{-1} dodecyl imidazole, 90 min. Bar = 0.5 μm.

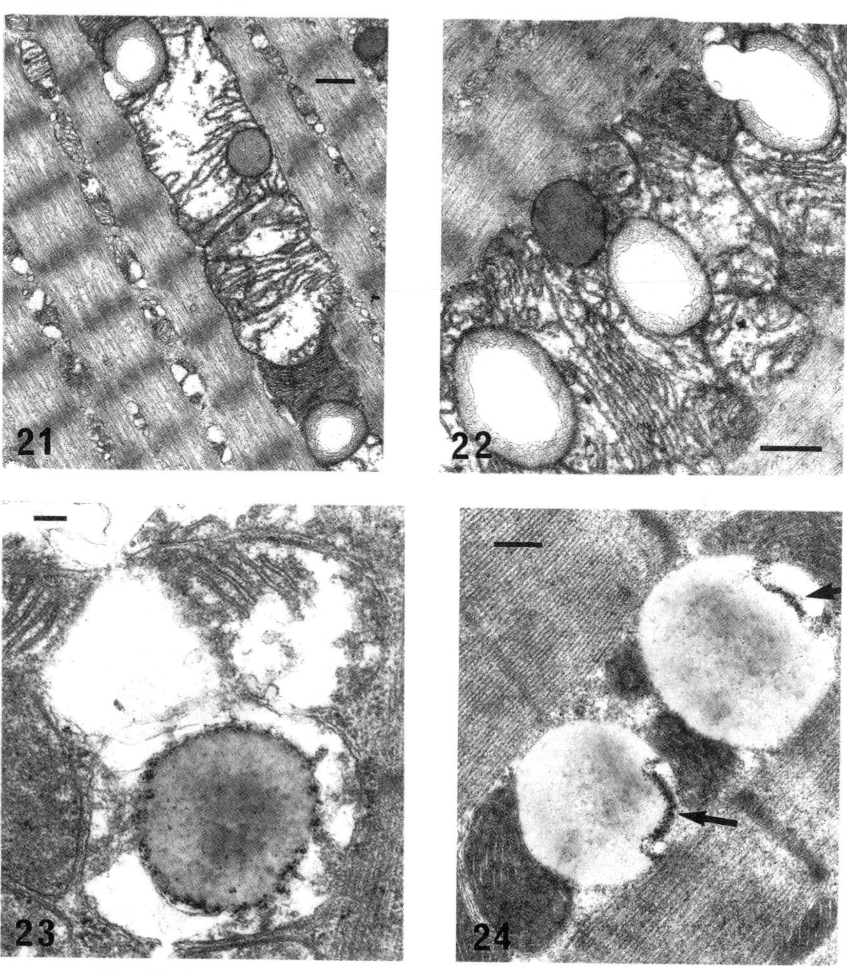

Fig. 21. Two lipid bodies forming by the breakdown of the centre of the lamellae. Damaged myofibrils. Mouse diaphragm; 50 μg ml^{-1} dodecyl imidazole, 90 min. Bar = 0.5 μm.

Fig. 22. Almost complete breakdown of central lamellar structures, forming lipid bodies among swollen mitochondria. Mouse diaphragm; 50 μg ml^{-1} dodecyl imidazole, 90 min. Bar = 0.5 μm.

Fig. 23. Formation of a lipid body within a damaged and swollen mitochondrion. Remnants of cristae which are attached to the lipid body are visible. Hamster diaphragm; high O_2 gassing, 90 min. Bar = 0.1 μm.

Fig. 24. Two complete lipid bodies between the mitochondria, possibly associated with T-tubules (arrows). Rat diaphragm; high O_2 gassing, 90 min. Bar = 0.2 μm.

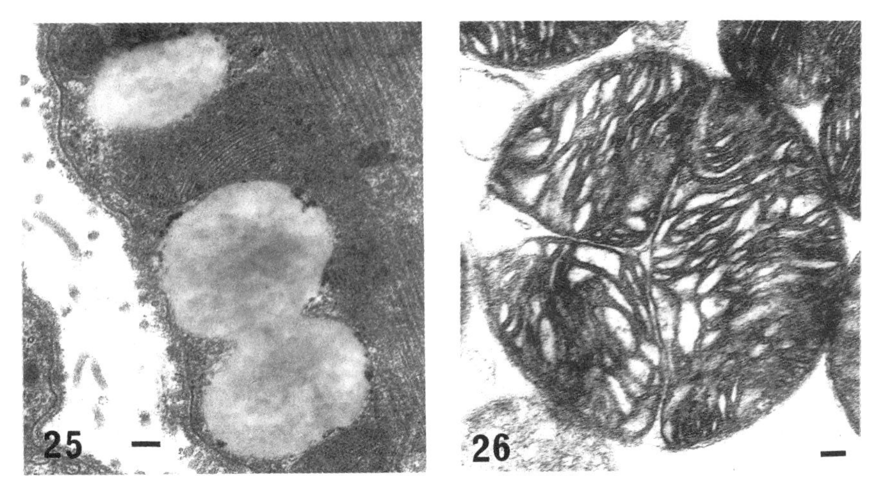

Fig. 25. Two complete lipid bodies lying between mitochondria and undergoing fusion. Rat diaphragm; high O_2 gassing; 90 min. Bar = 0.1 μm.

Fig. 26. Mitochondria subdivided with marked changes in the organisation of the cristae. Rat cardiomyocytes, 30 μg ml^{-1} dodecyl imidazole. Bar = 0.1 μm.

centre of the organelle (Figs 8, 20–22) whereas in others the lipid body develops at the periphery of the original mitochondrion (Figs 14–18). Finally, the membranous origins are lost (Figs 1, 13, 23–24) and the fully-formed lipid bodies sometimes fuse together (Fig. 25). Lipid bodies were also found in the cells of damaged rat (Figs 24 and 25) and hamster (Fig. 23) diaphragms so the formation of myelin figures and lipid bodies is widespread in rodent muscle cells.

Discussion

The rapid cellular damage in mammalian skeletal and cardiac muscle that is initiated by a variety of agents that are believed to raise $[Ca^{2+}]_i$ is usually accompanied by major changes in the ultrastructure of the mitochondria which swell and undergo changes in the organisation of their cristae, including the development of mitochondrial bars and internal septa (Duncan, 1988). The cristae are frequently changed into the electron-dense mitochondrial bars. Exceptionally, these changes in the organisation of the cristae result in the formation of myelin-like figures within the organelle. We conclude therefore that myelin-like figures develop within mitochondria of damaged muscle (Duncan, 1989). Membrane-bound lipid bodies are another characteristic and conspicuous feature of damaged muscle cells, developing rapidly with a time-course of 20–60 min, and the electronmicrographs presented here suggest strongly that they develop within mitochondria and that the lipid material is formed by the breakdown of their disorganised membranes. These may be the lamellae of either myelin-like figures or modified cristal membranes, tightly packed concentrically. It is evident that the membrane organisation is very plastic in the mitochondria of damaged muscle cells. Lipid bodies are particularly common in skeletal muscle cells in which cellular damage has been triggered by treatments that are believed to raise $[Ca^{2+}]_i$, such as anoxia (Fig. 13) or A23187 (Duncan, 1988).

Role of the lysosomal apparatus

A number of reports describe lipid bodies and myelin figures in muscle cells as being the products of lysosomal digestion, particularly during conditions of cell damage, for example following a vitamin E-deficient diet (Howes, Price & Blumberg, 1964) or in dystrophic muscle (Christie & Stoward, 1977), in ageing cardiac muscle (Topping & Travis, 1974) and in diseased and ageing human muscle (Mastaglia & Walton, 1982; Dubowitz, 1985; Walton, 1988). However, the status of the lysosomal system in muscle cells and its role in degradation is uncertain; vertebrate skeletal muscle appears to contain few typical lysosomes but macromolecules can enter by endocytosis, mainly at the T-tubule membrane. An increased rate of endocytosis occurs in such damage conditions as dystrophy, chronic denervation and degenerative conditions and it has been suggested that endocytosis is functionally related to increased lysosomal activity (Libelius *et al.*, 1978) and that lysosomes are part of the sarcotubular system (Seiden, 1973). Exposure of mammalian skeletal muscles to the lysosomotropic agents leucine methylester or dodecyl imidazole caused an initial swelling of the SR followed by its collapse which was accompanied by widespread and characteristic damage of the myofilament apparatus. However, this damage was not ameliorated by inhibitors of lysosomal cathepsins and it was concluded that acid hydrolases were not implicated (Duncan & Rudge, 1988; Shamsadeen & Duncan, 1989). In the present study both leucine methylester and dodecyl imidazole promoted mitochondrial swelling, severe changes in the organisation of the cristae and the formation of myelin figures and lipid bodies, but since these changes also (i) occur when the muscles are incubated with the inhibitors of acid hydrolases (Shamsadeen & Duncan, 1989), (ii) occur rapidly, and (iii) can be triggered by other agents such as A23187, we conclude that they are not necessarily the result of lysosomal digestion but rather one of the consequences of the sequence of biochemical events associated with cellular damage.

Calcium and phospholipase A_2

A number of reports now show that $[Ca^{2+}]_i$ measured directly rises in damaged skeletal muscle cells (Lopez *et al.*, 1985; Turner *et al.*, 1988; Mongini *et al.*, 1988) and Ca^{2+}-uptake by the mitochondria will be accelerated under these conditions, a process that is known to induce confirmational transitions (Hunter, Haworth & Southard, 1976). It is possible that the Ca^{2+}-dependent mitochondrial phospholipase A_2 (PLA_2) will be activated when the local $[Ca^{2+}]_i$ is elevated under damaging conditions (Saris & Van den Bosch, 1988), thereby rapidly promoting the turnover of membrane phospholipids, membrane breakdown and changes in mitochondrial permeability. Certainly, the formation of myelin figures described above appears to be related to a major reorganisation of the cristal membranes, and the lipid bodies are apparently a store of phospholipids and proteins resulting from the degradation of these membranes.

Alternatively, the mitochondrion is known to produce superoxides at two points in the electron transport system. Could excess production of these active species, stimulated by Ca^{2+}-uptake, swamp the mitochondrial superoxide dismutase and bring about the degradation of the cristal membranes?

Pulmonary surfactant

The formation of myelin figures and lipid bodies has features in common with the production of pulmonary surfactant that is synthesised in Type II alveolar epithelial cells of the mature mammalian lung. Pulmonary surfactant is a complex lipoprotein (10% protein, 76% phospholipid and 14% neutral lipid) that reduces alveolar surface tension during expiration. Phosphatidylcholine constitutes 80% of the phospholipid and the surfactant functions by forming a monomolecular layer at the gas–liquid interface, thereby reducing surface tension and preventing alveolar collapse (Corbet, 1984). The developmental sequence is believed to be as follows. The surfactant is assembled in the endoplasmic reticulum of the specialised Type II cells (granular pneumocytes), protein in the ribosomes and phospholipids at the membrane. Newly-synthesised phosphatidylcholine then translocates through the Golgi zone and appears in small lamellar bodies, while the protein moves through the Golgi zone and appears in the form of multivesicular bodies. These two types of particles fuse to form new lamellar bodies which, with the addition of more particles, develop into mature lamellar bodies which consist of concentric layers of phospholipid membrane. Lamellar bodies are extruded by exocytosis, whereupon the concentric layers unfold and produce long strings of material arranged in a lattice, called tubular myelin. Conversion of lamellar bodies into lattice myelin is thought to be Ca^{2+}-dependent (Corbet, 1984).

However, inspection of electronmicrographs which show the formation of these inclusion bodies in Type II cells (Goldfischer, Kikkawa & Hoffman, 1968; Kikkawa & Spitzer, 1969; Chevalier & Collet, 1973) reveals (i) a close similarity between mature lamellar bodies and the myelin figures of damaged muscle, particularly those shown in Figs 9–12, (ii) many developing lamellar bodies resemble damaged mitochondria as in Fig. 26 (see also Publicover *et al.*, 1977; 1979), (iii) some inclusion bodies in Type II cells are apparently enclosed within double membranes thereby resembling mitochondria, (iv) remnants of cristal membrane organisation can be detected in developing lamellar bodies of Type II cells, (v) many lamellar bodies in Type II cells show the membrane whorls breaking down apparently into lipid, thereby resembling Fig. 15, and (vi) occasionally in the present study, lipid bodies were seen being extruded through the sarcolemma of skeletal muscle with the released lipid forming a lattice-like mesh.

In view of these similarities it is tempting to speculate that the major changes in mitochondrial organisation and ultrastructure seen during muscle cell damage may provide a clue for the formation of inclusion bodies, lamellar bodies and lung surfactant, suggesting that these may be formed within organelles derived from modified mitochondria whose breakdown can be regulated in response to external stimuli. Surfactant secretion in Type II cells is triggered by the entry of extracellular Ca^{2+} (as with other forms of exocytosis), by the cellular level of cAMP and by phorbol myristate acetate (Corbet, 1984), suggesting that the protein kinase A and the protein kinase C pathways interact with levels of $[Ca^{2+}]_i$, perhaps modulating Ca^{2+} uptake by the mitochondria and thereby initiating the breakdown and modification of internal membranes via PLA_2 activity and so generating surfactant production.

Acknowledgement

The help of Miss S. Scott in preparing this manuscript is gratefully acknowledged.

References

Caulfield, J.B. (1966). Electron microscopic observations on the dystrophic hamster muscle. *Annals of the New York Academy of Sciences* **138**, 151–9.

Chevalier, G. & Collet, A.J. (1973). *In vivo* incorporation of choline-^{3}H, leucine-^{3}H and galactose-^{3}H in alveolar type II pneumocytes in relation to surfactant synthesis. A quantitative radioautographic study in mouse by electron microscopy. *Anatomical Record* **174**, 289–310.

Christie, K.N. & Stoward, P.J. (1977). A cytochemical study of acid phosphatases in dystrophic hamster muscle. *Journal of Ultrastructure Research* **58**, 219–34.

Corbet, A. (1984). Surfactant secretion and turnover. In *Hyaline Membrane Disease*, ed. L. Stern, pp. 35–61. New York: Grune & Stratton Inc.

Dubowitz, V. (1985). *Muscle Biopsy. A practical approach*. London: Bailliere Tindall.

Duncan, C.J. (1987). Role of calcium in triggering rapid ultrastructural damage in muscle: a study with chemically skinned fibres. *Journal of Cell Science* **87**, 581–94.

Duncan, C.J. (1988). Mitochondrial division in animal cells. In *The Division and Segregation of Organelles*, ed. S.A. Boffey & D. Lloyd, pp. 95–113, Society for Experimental Biology. Cambridge University Press.

Duncan, C.J. (1989). The mechanisms that produce rapid and specific damage to the myofilaments of amphibian skeletal muscle. *Muscle and Nerve* **12**, 210–18.

Duncan, C.J., Greenaway, H.C. & Smith, J.L. (1980). 2,4-dinitrophenol, lysosomal breakdown and rapid myofilament degradation in vertebrate skeletal muscle. *Naunyn-Schmiedeberg's Arch. Pharmacol.* **315**, 77–82.

Duncan, C.J. & Rudge, M.F. (1988). Are lysosomal enzymes involved in rapid damage in vertebrate muscle cells? A study of the separate pathways leading to cellular damage. *Cell and Tissue Research* **253**, 447–55.

Goldfischer, S., Kikkawa, Y. & Hoffman, L. (1968). The demonstration of acid hydrolase activities in the inclusion bodies of type II alveolar cells and other lysosomes in the rabbit lung. *Journal of Histochemistry and Cytochemistry* **16**, 102–9.

Howes, E.L. Jr, Price, H.M. & Blumberg, J.M. (1964). The effects of a diet producing lipochrome pigment (ceroid) on the ultrastructure of skeletal muscle in the rat. *American Journal of Pathology* **45**, 599–631.

Hunter, D.R., Haworth, R.A. & Southard, J.H. (1976). Relationship between configuration, function and permeability in calcium-treated mitochondria. *Journal of Biological Chemistry* **251**, 5069–77.

Kikkawa, Y. & Spitzer, R. (1969). Inclusion bodies of type II alveolar cells: species differences and morphogensis. *Anatomical Record* **163**, 525–42.

Libelius, R., Lundquist, I., Templeton, W. & Thesleff, S. (1978). Intracellular uptake and degradation of extracellular tracers in mouse skeletal muscle in vitro: the effect of denervation. *Neuroscience* **3**, 641–7.

Lopez, J.R., Alamo, L., Caputo, C., Wikinski, J. & Ledezma, D. (1985). Intracellular ionized calcium concentration in muscles from humans with malignant hyperthermia. *Muscle and Nerve* **8**, 355–8.

Mastaglia, F.L. & Walton, J. (1982). *Skeletal Muscle Pathology*. Churchill Livingstone.

Mongini, T., Ghigo, D., Doriguzzi, C., Bussolino, F., Pescarmona, G., Pollo, B., Schiffer, D. & Bosia, A. (1988). Free cytoplasmic Ca^{++} at rest and after cholinergic stimulus is increased in cultured muscle cells from Duchenne muscular dystrophy patients. *Neurology* **38**, 476–80.

Nag, A.C. & Cheng, M. (1982). Differentiation of fibre types in an extraocular muscle of the rat. *Journal of Embryology and Experimental Morphology* **71**, 171–91.

Publicover, S.J., Duncan, C.J. & Smith, J.L. (1977). Ultrastructural changes in muscle mitochondria in situ, including the apparent development of internal septa, associated with the uptake and release of calcium. *Cell and Tissue Research* **185**, 373–85.

Publicover, S.J., Duncan, C.J. & Smith, J.L. (1978). The use of A23187 to demonstrate the role of intracellular calcium in causing ultrastructural damage in mammalian muscle. *Journal of Neuropathology and Experimental Neurology* **37**, 544–57.

Publicover, S.J., Duncan, C.J., Smith, J.L. & Greenaway, H.C. (1979). Stimulation of septation in mitochondria by diamide, a thiol oxidising agent. *Cell and Tissue Research* **203**, 291–300.

Saris, N.-E.L. & Van den Bosch, H. (1988). Interaction of Sr^{2+} with Ca^{2+} release in mitochondria. *Journal of Bioenergetics and Biomembranes* **20**, 749–57.

Seiden, D. (1973). Effects of colchicine on myofilament arrangement and the lysosomal system in skeletal muscle. *Zeitschrift für Zellforschungs* **144**, 467–73.

Shamsadeen, N. & Duncan, C.J. (1989). Cytotoxic action of the lysosomotropic agent L-leucine methylester on mammalian skeletal muscle. The role of the sarcoplasmic reticulum in producing myofilament damage. *Virch. Arch. B. Cell. Path.* **57**, 315–21.

Stenger, R.J., Spiro, D., Scully, R.E. & Shannon, J.M. (1962). Ultrastructural and physiologic alterations in ischemic skeletal muscle. *American Journal of Pathology* **40**, 1–20.

Threadgold, L.T. (1976). *The Ultrastructure of the Animal Cell.* Pergamon International Library.

Tomanek, R.J. (1976). Ultrastructural differentiation of skeletal muscle fibers and their diversity. *Journal of Ultrastructure Research* **55**, 212–27.

Topping, T.M. & Travis, D.F. (1974). An electron cytochemical study of mechanisms of lysosomal activity in the rat left ventricular mural myocardium. *Journal of Ultrastructure Reearch* **46**, 1–22.

Turner, P.R., Westwood, T., Regen, C.M. & Steinhardt, R.A. (1988). Increased protein degradation results from elevated free calcium levels found in muscle from *mdx* mice. *Nature* **335**, 735–8.

Walton, J. (1988). *Disorders of Voluntary Muscle*. Churchill Livingstone.

JON D. GOWER, LISA A. COTTERILL
and COLIN J. GREEN

The importance of oxygen free radicals, iron and calcium in renal ischaemia

Introduction

Ischaemia encompasses a wide range of clinical conditions and is also an integral part of many surgical techniques, in particular transplantation. Organ retrieval usually involves a short period of warm ischaemia (WI) between cessation of the blood supply and harvesting the organ from the donor. This is followed by a much longer period of cold ischaemia (CI) in which the organ is flushed with and suspended in a cold asanguinous solution for transport to the recipient. The organs are then rapidly reperfused with fully oxygenated blood as soon as the vascular pedicle is reconstructed. Cooling depresses metabolism and very much slows the deterioration of ischaemic organs. However, some organs are particularly susceptible to ischaemic damage and it is currently considered inadvisable to store liver, heart or lungs for longer than 4 hr. Kidneys are usually stored for about 24 hr but storage periods up to 72 hr are not uncommon. There is no definitive safe storage time but rather the longer the period of ischaemia, the less chance there is of an organ functioning immediately upon transplantation. Acute renal failure may occur in transplanted kidneys which become enlarged with a pale cortex and a dark congested medulla and have a drastically impaired excretory capacity. Vascular injury is another possible complication in ischaemically damaged kidneys which are slow to perfuse when revascularised and develop a microcoagulopathy which results in an outflow block and venous stasis.

Many biological changes associated with ischaemia and reperfusion have been reported. These include depletion of high-energy adenine nucleotides (Calman, Quin & Bell, 1973), accumulation of metabolites such as H^+ ions leading to a significant fall in intracellular pH (Sehr *et al.*, 1979), release of lysosomal enzymes (Pavlock *et al.*, 1984), loss of membrane phospholipids (Southard *et al.*, 1984), and impaired function of intracellular organelles such as mitochondria (Arnold *et al.*, 1985) and endoplasmic reticulum (Schieppati *et al.*, 1985). However, the mechanisms underlying the pathology of organ deterioration due to ischaemia have yet to be conclusively elucidated.

The observation that many pathological changes only become evident after restoration of the blood supply to an ischaemic organ has led to the term 'reperfusion injury'. There is considerable evidence that an important part of reperfusion injury is oxidative damage initiated by the incoming molecular oxygen. Single-electron reduction of O_2 leads to the formation of superoxide radicals ($O_2^{\cdot-}$); further reduction yields H_2O_2 and the highly reactive hydroxyl radical ($^{\cdot}OH$). Under normal physiological circumstances, production of these reactive species is low because O_2 metabolism is carefully controlled by enzymes such as cytochrome oxidase which catalyses the $4e^-$ reduction of O_2 directly to H_2O. Cells contain a number of protective enzymes such as superoxide dismutase which converts $O_2^{\cdot-}$ to H_2O_2 and catalase and peroxidases which reduce H_2O_2 to H_2O. There are also a number of smaller molecules with important biological antioxidant activity such as the lipid-soluble vitamin E, ascorbic acid and glutathione. Although these systems ensure that oxidative damage is kept to a minimum under normal circumstances, this does not appear to be the case when organs are reperfused following periods of ischaemia.

Oxygen-derived free radicals have been implicated in reperfusion damage to many tissues (e.g. Schoenberg *et al.*, 1983; McCord, 1985; Bolli, 1988) including the kidney (Paller, Hoidal & Ferris, 1984; Laurent & Ardaillou, 1986; Ratych & Bulkley, 1986) and in the 'storage-damage syndrome' (Koyama *et al.*, 1985; Fuller, Gower & Green, 1988). One damaging free-radical-mediated process is the peroxidation of polyunsaturated fatty acids (Wills, 1969), a chain reaction which can result in extensive membrane damage.

We sought evidence for the involvement of oxygen-derived free radicals in renal injury by assaying for markers of lipid peroxidation in homogenates prepared from kidneys which had been subjected to periods of ischaemia. Periods of both warm (Green *et al.*, 1986*c*) and cold ischaemia (Green *et al.*, 1986*a*) were found to increase significantly the rate of formation of lipid peroxidation markers during subsequent incubation at 37 °C *in vitro*. In the CI experiments, kidneys were either stored in a poor storage medium (isotonic saline) for 24 hr or for periods up to 72 hr in the more efficacious and clinically-approved hypertonic citrate solution (HCA) developed by Ross, Marshall & Escott (1976). The results showed that there was a good correlation between the formation of markers of lipid peroxidation measured *in vitro* and the physiological dysfunction of the stored organs upon transplantation *in vivo*. Reperfusion of the ischaemic organs with oxygenated blood *in vivo* generally led to further rises in the extent of lipid peroxidation in these organs. Addition of free-radical scavengers (mannitol and uric acid) or the iron-chelator desferrioxamine to the flush and storage solutions significantly inhibited the adverse rises in lipid peroxidation products following the ischaemic period (Green *et al.*, 1986*b*).

These initial findings provided good circumstantial evidence for a role of oxygen-derived free radicals in post-ischaemic oxidative damage to the kidney. This raised the question of why a period of ischaemia should compromise an organ in this way. One possibility is that during the ischaemic period there is a reduction in the level of antioxidant defenses. Another possibility is that changes occur during ischaemia which result in a burst of oxidant production upon reoxygenation. One such effect is the conversion of the enzyme xanthine dehydrogenase to xanthine oxidase during ischaemia (Roy & McCord, 1983), possibly as a result of a calcium-dependent proteolysis. Upon reoxygenation this enzyme utilises hypoxanthine, which has accumulated due to catabolism of ATP, and forms $O_2^{\cdot-}$ radicals from the incoming oxygen. Mitochondrial injury and a build-up of reduced components of the electron transport chain during ischaemia may also increase $O_2^{\cdot-}$ production on reoxygenation due to increased leakage of single electrons onto O_2. In addition, accumulation of polymorphonucleocytes due to the release of chemotactic factors from ischaemic tissues may cause endothelial injury through the extracellular production of free radicals derived from the respiratory burst (Granger *et al.*, 1989).

Two metal ions, iron and calcium, are important determinants of free-radical-mediated processes, cell injury and vascular disturbance, and their possible role in the pathology of ischaemic/reperfusion damage in the kidney is now discussed in detail. Particular emphasis is placed on the temporal sequence of events, special attention being directed towards early changes which may occur during the ischaemic period itself and which may therefore be important in mediating the extent of subsequent reperfusion injury.

The role of iron

The importance of transition metals in catalysing damaging free-radical-mediated processes has long been recognised. Of particular biological relevance is the iron-catalysed formation of hydroxyl radicals ($^{\cdot}OH$) from less reactive precursors ($O_2^{\cdot-}$ and H_2O_2) via the metal-catalysed Haber–Weiss reaction (Halliwell, 1978). The highly reactive $^{\cdot}OH$ radical can damage all types of biological macromolecule and initiate lipid peroxidation (Gutteridge, 1984). Other studies suggest that some iron-centred species may themselves catalyse hydrogen abstraction from polyunsaturated fatty acids and hence directly initiate lipid peroxidation (Minotti & Aust, 1987). Furthermore, iron salts are known to decompose lipid hydroperoxides to reactive peroxy and alkoxy radicals which can attack further molecules of polyunsaturated fatty acids and hence propagate the chain reaction of lipid peroxidation (Halliwell & Gutteridge, 1984). Some low molecular weight chelates of iron such as ATP and EDTA increase the reactivity of the metal (Dunford, 1987) whereas some high molecular weight chelators such as desferrioxamine (DFX) bind iron with high affinity (10^{31}) (Keberle, 1964) and prevent the metal ion from catalysing adverse reactions (Gutteridge, Richmond & Halliwell, 1979).

Administration of DFX (i.v., 15 mg/kg) to rabbits 15 min before reperfusion of kidneys which had been subjected to 60 or 120 min of warm ischaemia was found to inhibit significantly the adverse rises in markers of lipid peroxidation (Green *et al.*, 1986*c*). DFX was also highly effective at reducing levels of oxidative membrane damage in kidneys which had been subjected to cold ischaemia by storage either in isotonic saline solution for 24 hr (Green *et al.*, 1986*b*) or for periods of up to 72 hr in HCA (Gower *et al.*, 1989*a*). The most effective regime proved to be i.v. administration of DFX both before the removal of kidneys for storage and before reperfusion of the autotransplanted organs. Analysis of markers of lipid peroxidation in different regions of the kidney revealed that DFX was particularly effective at inhibiting lipid peroxidation in the cortex following ischaemia; whereas in the medulla, which contains relatively high levels of cyclooxygenase (Robak & Sobanska, 1976), a less marked decrease in lipid peroxidation was observed upon DFX treatment.

These results strongly suggested a role for iron in mediating oxidative membrane damage which is observed on reperfusion. Under normal physiological circumstances, the overwhelming majority of iron in the body is stored in 'safe' sites which prevent transition of its redox state and hence catalysis of damaging reactions involving single electrons. These 'safe' sites include haemoglobin and transferrin in the extracellular medium, and ferritin, a predominantly intracellular protein which stores up to 4500 atoms of iron in the form of ferric hydroxides (Aisen & Listowsky, 1980). There is, however, growing evidence for a small pool of intracellular iron which is chelated to low molecular weight (LMW) species such as ATP, citrate and glycine (Bakkeren *et al.*, 1985; Mulligan, Althaus & Linder, 1986). This pool may represent iron which is in transit from the extracellular milieu (transferrin) to intracellular storage sites (ferritin) or iron which is required for synthetic purposes. It is likely that this pool of LMW iron, though small, has the potential for exerting some degree of oxidative stress on the cell.

We hypothesised that ischaemia may result in altered intracellular iron homeostasis leading to an increase in the level of catalytic forms of the metal. Evidence for this was sought by developing a method for quantitating the amount of intracellular iron available for chelation by DFX (Gower, Healing & Green, 1989*b*). Low speed supernatants of tissue homogenates were incubated with an excess of DFX for 60 min and the parent drug and its iron-bound form, ferrioxamine (FX), were extracted using solid-phase extraction cartridges and analysed by reversed-phase HPLC with dual-wavelength detection (FX was detected at 430 nm and DFX at 226 nm). Standard curves obtained after known amounts of iron were incubated with DFX demonstrated that, with suitable precautions, the ratio of the area of the FX peak/area of DFX peak gave an accurate determination of DFX-available (DFX-A) iron levels in the 1–25 μM range. Using this method, DFX-A iron levels were determined in kidneys subjected to periods of warm and cold ischaemia and subsequently reperfused *ex vivo* with an oxygenated medium on an isolated perfusion apparatus at 37 °C.

There were measurable levels of DFX-A iron present in both cortex and medulla of fresh control kidneys prior to any ischaemic insult (Figs 1 and 2). Following 2 hr warm ischaemia or 24 hr cold storage in HCA, levels of DFX-A iron significantly increased by about two-fold in both the cortex and medulla (Gower, Healing & Green, 1989*c*). Storage in HCA for longer periods (up to 72 hr) resulted in levels of DFX-A iron which were generally higher, but not significantly so, than after the less damaging 24 hr period of cold ischaemia.

Much more dramatic differences were observed in the levels of DFX-A iron between the ischaemic groups when the stored organs were reperfused (Healing *et al.*, 1990). After 24 hr CI, the levels of DFX-A iron immediately decreased upon reperfusion and returned rapidly to control levels in both the cortex and medulla within 5 min (Figs 1 and 2). In contrast, following the more physiologically damaging 48 hr period of CI, DFX-A iron levels remained elevated in both regions of the kidney during the first 5 min of reperfusion and returned to control levels only after 30 min. A similar response was observed in the medulla of kidneys when the ischaemic time was increased from 48 to 72 hr (Fig. 2); however, in the cortex of these organs, DFX-A iron levels actually increased during the first 5 min of reperfusion before returning to control levels after 30 min. There were no significant differences in the total iron content of cortex and medulla between any of the groups as measured by atomic

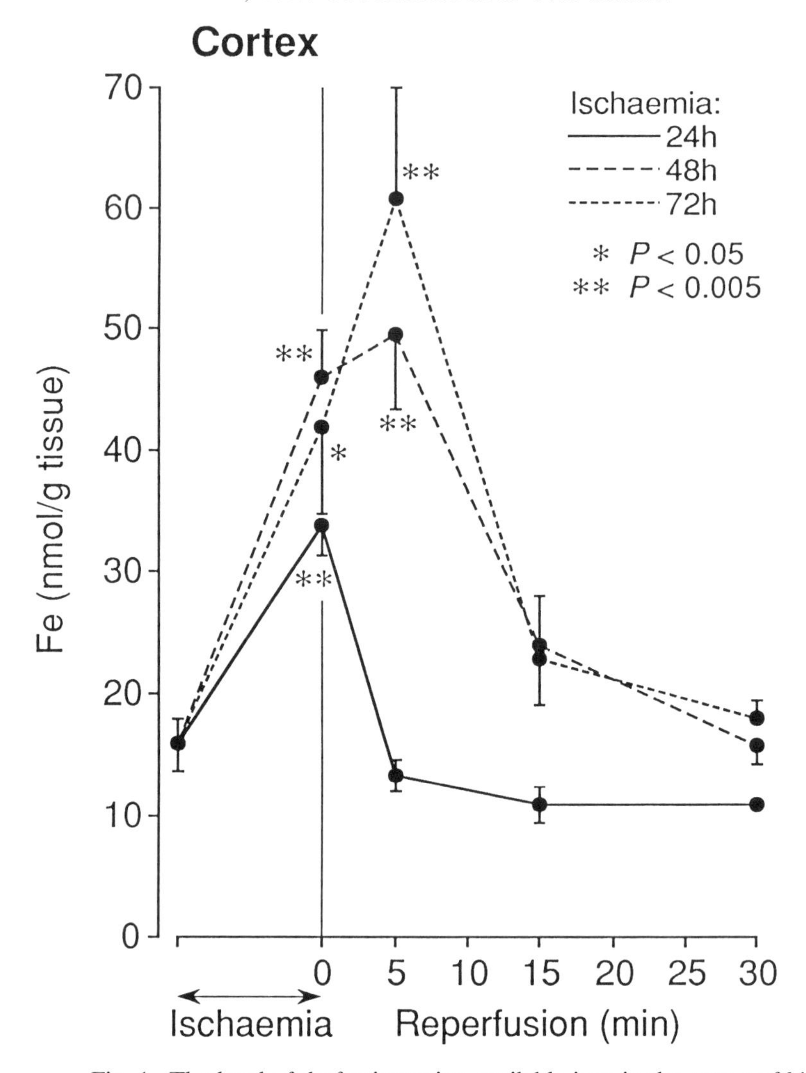

Fig. 1. The level of desferrioxamine-available iron in the cortex of kidneys subjected to 24, 48 or 72 hr cold ischaemia followed by up to 30 min *ex vivo* reperfusion at 37 °C with an oxygenated asanguinous perfusate. Values represent the mean ± SEM of six separate determinations performed in duplicate.

absorption spectroscopy. Thus, these results clearly showed that both warm and cold ischaemia led to a redistribution of intracellular iron to forms more available for chelation by DFX. In view of our earlier findings that DFX was more effective at inhibiting increased levels of lipid

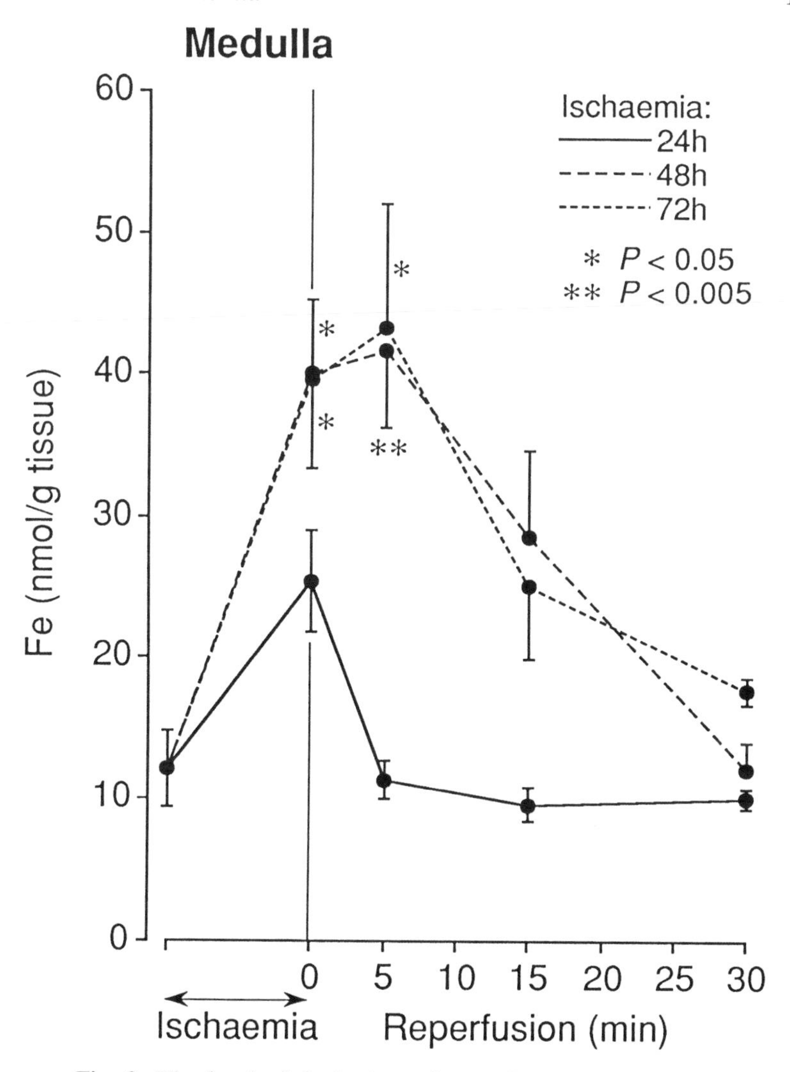

Fig. 2. The level of desferrioxamine-available iron in the medulla of kidneys subjected to 24, 48 or 72 hr cold ischaemia followed by up to 30 min *ex vivo* reperfusion at 37 °C with an oxygenated asanguinous perfusate. Values represent the mean ± SEM of six separate determinations performed in duplicate.

peroxidation in the cortex following ischaemia (Gower *et al.*, 1989*a*), it is interesting to note that increases in DFX-A iron levels during ischaemia were more pronounced in this region compared to the medulla.

It is likely, though not proven, that the source of increased levels of

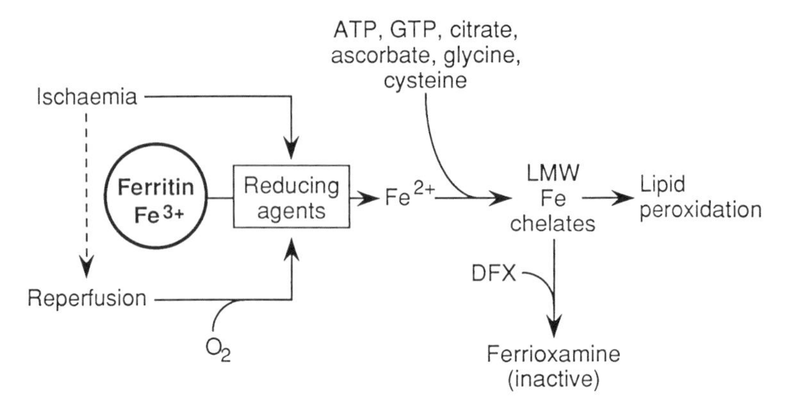

Fig. 3. A schematic diagram depicting the release of iron from ferritin by reducing agents produced during ischaemia and reperfusion, the formation of low molecular weight (LMW) iron chelates and the subsequent catalysis of lipid peroxidation which is prevented by chelation of catalytic iron complexes with desferrioxamine (DFX).

DFX-A iron during ischaemia was ferritin. Although surprisingly little is known about the release of iron from ferritin, it is thought to involve the reduction of Fe^{3+} to Fe^{2+} (Funk *et al.*, 1985). This reduction could be facilitated by the low oxygen tension encountered as a result of ischaemia and by the accompanying fall in pH. Such an environment would favour increased levels of reducing agents which may release iron from this protein (Fig. 3). Exogenous redox-active quinones such as adriamycin have been shown to catalyse lipid peroxidation in the presence of ferritin especially under hypoxic conditions (Vile & Winterbourn, 1988). However, the identity of possible endogenous agents capable of performing this function and produced as a result of ischaemia has yet to be established. Experiments *in vitro* have demonstrated that lipid peroxidation in microsomes is stimulated in the presence of purified ferritin and flavin mononucleotide during aerobic incubation following a period of hypoxia (Goddard *et al.*, 1986).

Further release of iron into the DFX-A pool on reperfusion of kidneys stored for longer periods may be the result of iron release from ferritin by reducing agents formed from the incoming oxygen (Fig. 3). Superoxide anions, which may be produced in significant amounts upon reperfusion, have been shown to release iron from ferritin *in vitro* (Biemond *et al.*, 1988; Monteiro & Winterbourn, 1988). Alternatively, other reaction products of superoxide or oxygen itself with as yet unidentified cellular components may play an important role. It is also possible that continued high levels of DFX-A iron during the reperfusion period were due to the impairment of uptake mechanisms, as yet poorly understood, of intracellular iron species into 'safe' sites.

These studies demonstrate that redistribution of intracellular iron to forms more accessible to DFX occurs as an early event during the ischaemic period itself. The fact that increased levels of iron become available for chelation by DFX makes it highly likely that iron redistribution is an important underlying cause of the increased oxidative damage which occurs upon reperfusion. However, extended and hence more damaging periods of cold ischaemia did not lead to correspondingly higher levels of delocalised iron immediately after the ischaemic period (Fig. 3). Rather, extended periods of elevated delocalised iron levels during the immediate reoxygenation period appeared to be more important in determining the degree of reperfusion damage to rabbit kidneys. Chelation of the increased levels of intracellular LMW iron complexes by agents such as DFX inhibit the initiation and propagation of damaging events such as lipid peroxidation (Fig. 3) and may therefore be a worthwhile approach to combating post-ischaemic renal failure.

The role of calcium

Calcium ions play a crucial role in the control of a wide variety of biological functions (Evered & Whelan, 1986). These include regulation of cell metabolism through the activation of a considerable number of enzymes and the maintenance of normal blood flow via controlled production of a range of eicosanoids. Careful maintenance of the gradient between low (10^{-7} M) cytosolic calcium levels and the high extracellular concentration (10^{-3} M) is therefore essential. Under normal physiological circumstances, energy-dependent pumps remove calcium from the cell and also sequester excess calcium into intracellular organelles (Carafoli, 1987). Calcium ions may enter the cell only through specific voltage- or receptor-operated channels (Carafoli, 1987) and intracellular calcium levels are also controlled by the receptor-mediated turnover of phosphatidylinositides which involves a GTP-binding protein and a specific phospholipase C (Berridge, 1984).

The rapid depletion of ATP levels upon ischaemia will compromise calcium pumps and a gain in cellular calcium levels is a consistent feature in tissues subjected to ischaemia and reperfusion (Nayler *et al.*, 1988). This is likely to lead to a derangement in cell function, and continued elevations in intracellular calcium may cause altered cell morphology and irreversible cell injury (Farber, 1981). There is considerable interest in the possible involvement of calcium in ischaemic/reperfusion damage (Cheung *et al.*, 1986; Opie, 1989). However, it is unclear as to whether altered intracellular calcium homeostasis is an important initiator of damage or whether it plays a much later role in the pathology of ischaemic/reperfusion injury.

We rendered rabbit kidneys cold ischaemic in storage solutions containing various agents which either affect calcium movements or interfere with calcium-dependent enzymes (Cotterill *et al.*, 1989*a*). Oxidative damage was subsequently assessed by measuring formation of markers of lipid peroxidation in tissue homogenates *in vitro*. Blockage of voltage-operated channels by verapamil reduced the extent of oxidative damage to low levels in both the cortex and medulla of kidneys following 24 hr storage in isotonic saline but had no effect on oxidative damage after more prolonged (72 hr) storage of organs in the superior HCA solution. Elevation of extracellular calcium levels by addition of $CaCl_2$ (1 mM) to the storage medium increased oxidative damage to significantly greater levels than in organs stored in saline alone but also had no effect when added to HCA. It was concluded that influx of extracellular calcium through voltage-operated channels (VOCs) was a significant mediator of oxidative damage to organs following storage in saline. In HCA, however, the adverse effect of extracellular calcium appeared to be prevented, or at least slowed. This may have been due to chelation of calcium by the excess of citrate (55 mM) in this medium or because the better ionic balance of this storage solution may have prevented activation of VOCs during the ischaemic period.

Storage of rabbit kidneys in the presence of A23187, an ionophore which permeabilises both plasma and intracellular membranes to calcium, resulted in post-ischaemic rates of lipid peroxidation which were significantly greater than the already elevated levels observed after storage in either saline for 24 hr or HCA for 72 hr (Cotterill *et al.*, 1989*a*). Ruthenium red, a polysaccharide dye which inhibits mitochondrial calcium transport, also potentiated oxidative damage to organs following CI in either of the two media. These results suggested that, even in the absence of extracellular calcium effects, redistribution of calcium within the cell could take place during the ischaemic period and contribute to increased peroxidation of cellular lipids upon reoxygenation.

There are several possible ways in which calcium ions may potentiate free-radical-mediated post-ischaemic injury. Conversion of xanthine dehydrogenase to xanthine oxidase, which may be an important source of $O_2^{\overline{\cdot}}$ radicals following ischaemia (Roy & McCord, 1983), can be catalysed by a calcium-dependent protease (McCord, 1985). In our cold ischaemic kidney model, inhibition of xanthine oxidase by addition of allupurinol to the saline storage medium partially prevented the increase in lipid peroxidation following storage in the presence of A23187 (Cotterill *et al.*, 1989*b*). Mitochondrial injury due to calcium overload of this organelle is often seen upon reperfusion (Arnold *et al.*, 1985) and may cause increased leakage of single electrons from the electron transport chain onto O_2, thus increasing O_2^{-} production. Another possible link between raised intracellular calcium concentrations and increased oxidative stress is the involvement of calcium-dependent phospholipases which hydrolyse membrane phospholipids, releasing free fatty acids (FFAs) and leaving residual lysophosphatides in the membrane. This possibility was borne out by the observation that specific inhibition of phospholipase A_2 by addition of dibucaine to the storage solution resulted in significant protection against oxidative membrane damage following ischaemia (Cotterill *et al.*, 1989*b*).

Evidence for phospholipase activation during CI was obtained by analysing free fatty acids in freeze-clamped renal tissue by gas liquid chromatography (Cotterill *et al.*, 1989*c*). Levels of unsaturated FFAs ($C_{18:1}$, $C_{18:2}$, $C_{20:4}$) were found to rise considerably in kidneys stored for 72 hr in the clinically approved HCA solution. No significant changes were observed in the levels of saturated ($C_{16:0}$, $C_{18:0}$) FFAs. The high concentration of free arachidonic acid ($C_{20:4}$) observed after 72 hr CI is likely to be of particular importance as release of this fatty acid from the membrane is the rate limiting step in the formation of prostaglandins (Isakson *et al.*, 1978). This may therefore lead to a derangement in the production of these vasoactive substances when the organ is reperfused with resulting important consequences in the endothelial lining of the vascular bed (Schlondorff & Ardaillon, 1986). Indeed, it has been demonstrated that following ischaemic insult there is a decrease in the level of the potent vasodilator prostacyclin and an elevation in the formation of thromboxanes which cause vasoconstriction (Lelcuk *et al.*, 1985; Schmitz *et al.*, 1985), a situation which may contribute to the 'no-reflow' phenomenon observed when some organs are transplanted. The release of free arachidonic acid during CI is likely to be the main cause of the increased rate of indomethacin-inhibitable peroxidation via the cyclooxygenase pathway which we have observed in the medulla of stored kidneys (Gower *et al.*, 1989*a*). In addition, increased phospholipase activity may result in the formation of lipoxygenase products such as leukotrienes which are powerful mediators of vascular shock and have been implicated in ischaemic injury (Lefer, 1985).

Storage of kidneys in the presence of dibucaine or A23187 demonstrated a good correlation between the extent of free fatty acid accumulation and the rate of post-ischaemic lipid peroxidation. Possible relationships between calcium, phospholipase activity and lipid peroxidation are shown in Fig. 4. Early redistribution of intracellular calcium during the ischaemic period leads to activation of phospholipases during the hypoxic period. The released unsaturated FFAs, unprotected by the membrane-bound antioxidant vitamin E, would form excellent targets for free-radical attack upon reoxygenation. The resulting alkoxy and peroxy radicals may then initiate peroxidation of membranes directly or may breakdown to relatively stable hydroperoxides which could diffuse to other sites in the cell and stimulate lipid peroxidation through interaction with catalytic iron complexes which regenerate reactive lipid radicals (Halliwell & Gutteridge, 1984). In addition, the concomitant build-up of residual lysophosphatides in the membrane, due to phospholipase activation, alter fluidity and permeability (Weltzem, 1979) and may render membranes more susceptible to free-radical attack (Ungemach, 1985). Peroxidation of membrane lipids results in a loss of the fatty acid content of membranes, and also increases lysophosphatide levels (Ungemach, 1985) and membrane rigidity (Demopoulos *et al.*, 1980). As phospholipase A_2 activity is higher in rigid membranes (Momchilova, Petkova & Koumanov, 1986), elevated rates of lipid peroxidation upon reoxygenation may lead to further increases in phospholipase activation, and the damaging cycle of events shown in Fig. 4 may therefore ensue. This would lead to extensive damage to membranes which would become permeabilised to calcium and hence cytosolic calcium levels would rise further. Oxidative damage has also been shown to inhibit the plasma membrane calcium-extruding system (Nicotera *et al.*, 1985) and loss of the ability of intracellular organelles to sequester calcium (Bellomo *et al.*, 1985). Highly elevated levels of cytosolic calcium resulting from the above sequence of events (Fig. 4) could alter cell morphology causing blebbing (Jewell *et al.*, 1982) and, in conjunction with free-radical-mediated damage, would lead to irreversible injury to the cell.

Recently we have investigated the possibility that ischaemia followed by reoxygenation may affect the cleavage of membrane-bound phosphatidylinositols (PIP_2) in the kidney. This secondary messenger system involves the formation of inositol triphosphate (IP_3) and diacylglycerol (DAG) (Berridge, 1984). IP_3 mobilises calcium from intracellular stores and DAG stimulates phosphorylation of protein kinase C (PKc), a process which requires phospholipids and calcium for maximum activity (Berridge, 1984). In the kidney, PIP_2 hydrolysis through activation of α_1-

---

Fig. 4. A scheme depicting the possible relationships between increased intracellular calcium (Ca) levels, phospholipase A_2 (PLA_2) activity, lipid peroxidation and membrane damage. Abbreviations: FFA, free fatty acids; LOO·, lipid peroxy radical; LOOH, lipid hydroperoxide; Fe, catalytic iron complex.

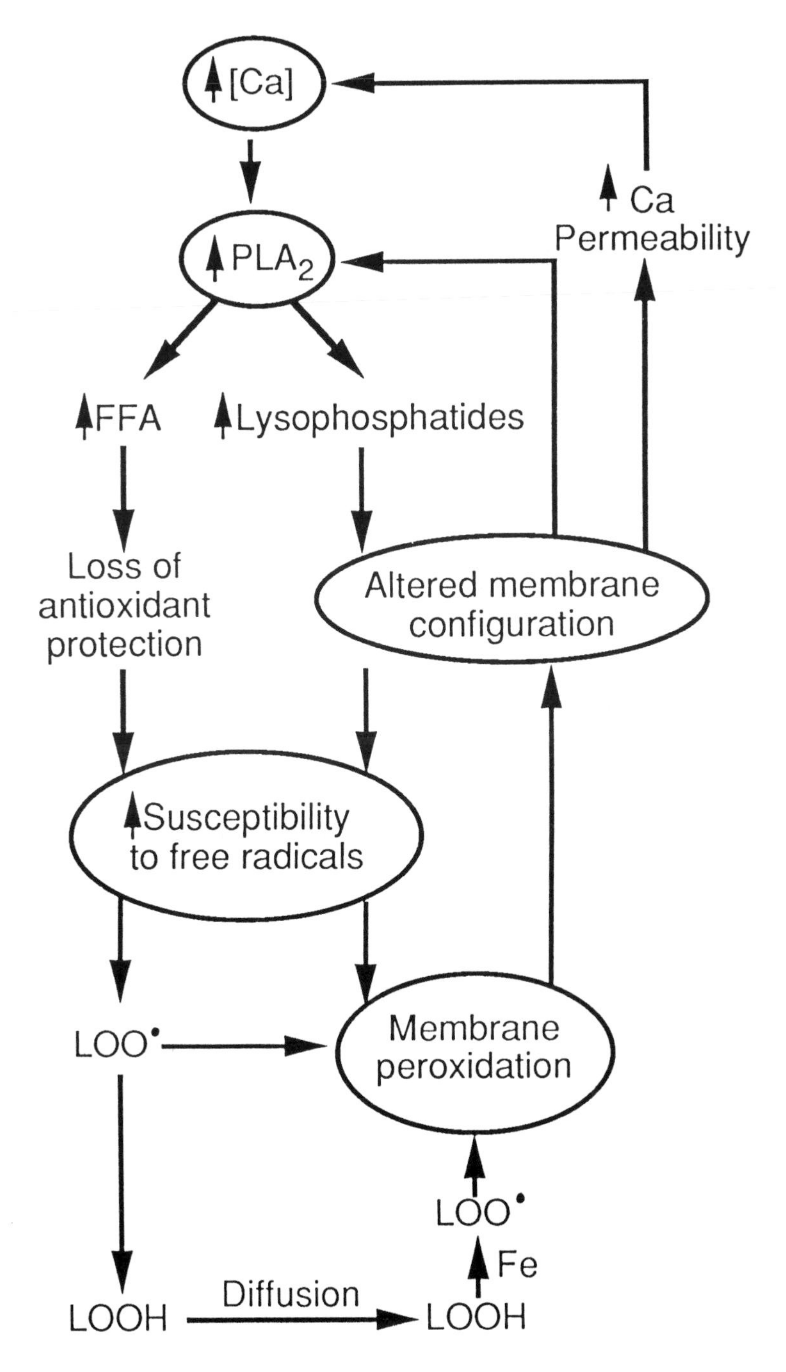
[Ca]
PLA2
Ca
Permeability
FFA
Lysophosphatides
Loss of
antioxidant
protection
Altered membrane
configuration
Susceptibility
to free radicals
LOO•
Membrane
peroxidation
LOO•
Fe
LOOH
Diffusion
LOOH

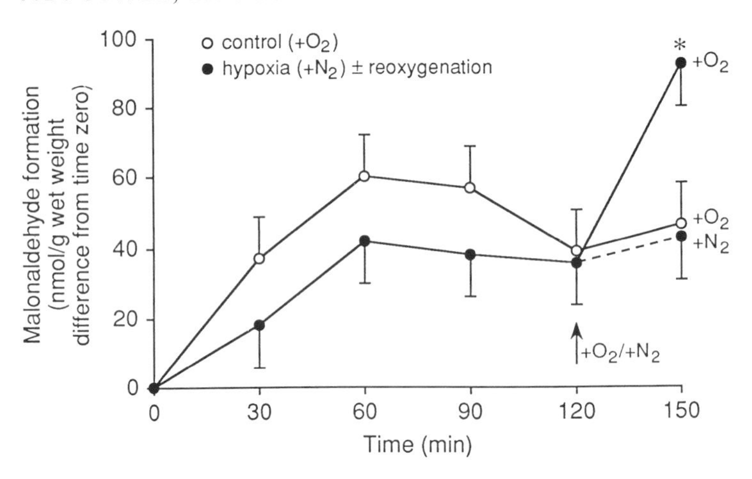

Fig. 5. Lipid peroxidation in kidney cortical slices incubated at 37 °C in Krebs–Ringer bicarbonate solution for 120 min under an atmosphere of N_2 (●) or O_2 (○) followed by regassing with either O_2 (———) or N_2 (– – – –) and incubation for a further 30 min. Values are the means ± 95% confidence limits of six separate determinations performed in duplicate. Lipid peroxidation was measured as thiobarbituric acid-reactive material and expressed as malonaldehyde formation. The asterisk (*) indicates $P<0.0001$ comparing pre-O_2 values with post-O_2 values.

adrenoceptors evokes a multiple response: increased sodium reabsorption (Hesse & Johns, 1984), prostanoid production and vasoconstriction (Cooper & Malik, 1985), gluconeogenesis (Kessar & Saggerson, 1980) and inhibition of renin release (Matsumura *et al.*, 1985).

In this set of experiments, kidney cortical slices were incubated *in vitro* at 37 °C either under an atmosphere of 95% O_2:5% CO_2 (control) or gassed with and incubated under N_2 (hypoxia). After 120 min all slices were then oxygenated and incubated aerobically for a further 30 min. The formation of lipid peroxidation products increased during the first 60 min of incubation in the presence of O_2 and then levelled off over the remaining period (Fig. 5). Lipid peroxidation was also evident in the hypoxic slices but proceeded at a slower rate than that of the oxygenated samples and also levelled off after 60 min (Fig. 5). Reoxygenation following 120 min of hypoxia resulted in a significant ($P<0.0001$) increase in the rate of lipid peroxidation which was not observed in the slices kept under N_2 or when slices incubated in the presence of oxygen for 120 min were regassed (Fig. 5). This *in vitro* model system therefore closely mimicked the increase in free-radical-mediated oxidative membrane damage which occurs when whole kidneys are subjected to ischaemia and reperfusion.

In order to determine the rate of PIP_2 hydrolysis under these conditions, radiolabel was incorporated into the membrane-bound phosphatidylinositol pool by incubating slices in the presence of myo-(2-^3H)-inositol for 60 min at 37 °C. The slices were then repeatedly washed and incubated under aerobic or hypoxic conditions for 120 min followed by oxygenation. Aliquots of slices were taken every 30 min and hydrolysis products of PIP_2 (IP_3 and its subsequent metabolites inositol bisphosphate (IP_2) and inositol monophosphate (IP_1)) were analysed by HPLC with scintillation counting (Irvine *et al.*, 1985). The extent of PIP_2 hydrolysis was also quantitated by a simple procedure using short Dowex anion-exchange columns (Berridge *et al.*, 1983). Samples were loaded onto the columns and two fractions were collected; fraction 1 containing inositol and glycerophosphoinositol was first eluted with 60 mM ammonium formate/5 mM sodium borate (12 ml) (fraction 1) and inositol phosphates (IP_1, IP_2 and IP_3) were then eluted with 1 M ammonium formate/0.1 M formic acid (12 ml) (fraction 2). The ratio of the radioactivity in fraction 2 to the radioactivity in fraction 1 was taken as a measure of PIP_2 breakdown.

There was no change in the rate of PIP_2 breakdown in the hypoxic slices during the 120 min incubation period compared with control slices incubated under aerobic conditions either in the presence of calcium or in calcium-free medium containing EGTA (Fig. 6). However, immediately upon reoxygenation of the hypoxic slices incubated in the presence of calcium, PIP_2 breakdown increased rapidly (Fig. 6). This increase was highly significant ($P=0.0002$) and was maintained over the remaining 30 min of aerobic incubation. No increase in PIP_2 breakdown was observed when the slices incubated in the presence of calcium and oxygen for 120 min were regassed with 95% O_2:5% CO_2 (Fig. 6). Furthermore, reoxygenation of hypoxic slices in calcium-free medium (+EGTA) had no effect on the rate of PIP_2 breakdown and no significant changes occurred in slices incubated in the presence of O_2 and EGTA (Fig. 6).

These results clearly demonstrated that hydrolysis of phosphatidylinositols to secondary messenger products is activated very rapidly upon reoxygenation of renal tissue following a period of hypoxia. Inhibition of the effect by EGTA strongly suggested the involvement of calcium. No changes in PIP_2 breakdown were observed during hypoxia itself. One of the products of lipid peroxidation, 4-hydroxynonenal has been shown to stimulate adenylate cyclase, guanylate cyclase and PIP_2 breakdown *in vitro* (Dianzani *et al.*, 1989). It is therefore possible that increased levels of aldehydic products of lipid peroxidation produced on reoxygenation were responsible for increased PIP_2 hydrolysis. In addition, both lipid peroxidation and high calcium-dependent phospholipase

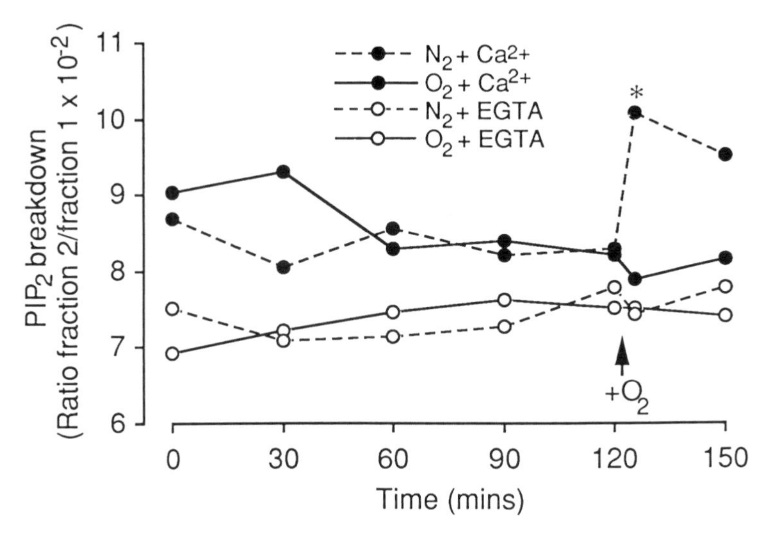

Fig. 6. Phosphatidylinositol breakdown in kidney cortical slices incubated at 37 °C in Krebs–Ringer bicarbonate solution containing calcium (2.5 mM) (●) or EGTA (10 mM) (○) for 120 min under an atmosphere of N_2 (– – – –) or O_2 (———) followed by reoxygenation and aerobic incubation for a further 30 min. Values represent the means of eight separate determinations performed in duplicate. PIP_2 breakdown was determined by anion-exchange chromatography (Berridge *et al.*, 1983) and expressed as the ratio of fraction 2 (inositol tris-, bis- and mono-phosphates)/fraction 1 (inositol and glycerophosphoinositol). The asterisk (*) indicates P=0.0002 comparing pre-O_2 values with post-O_2 values.

A_2 activity alter membrane configuration (Ungemach, 1985) and this may affect the interaction of phospholipase C with membrane-bound regulatory components or make it more accessible to its substrate (PIP_2). It is also possible that highly altered calcium levels during the immediate reperfusion period were sufficient to evoke a response of the phospholipase C system. Rapid hydrolysis of PIP_2 on reoxygenation following ischaemia would lead to deregulation of receptor-mediated function through this intracellular secondary messenger system. One of the consequences of this is likely to be an imbalance in eicosanoid production and vascular disturbances which may contribute to the pathogenesis of renal damage which occurs under these conditions.

Ischaemia	Reperfusion
↓ ATP Accumulation of Metabolites eg. Hypoxanthine ↓ pH ↑ LMW Iron Ca redistribution	↑ Oxygen Free Radicals ↑ Lipid Peroxidation Imbalance in Prostaglandin/ Thromboxane production Production of Inflammatory Mediators eg. Leukotrienes
Reversible Cell Injury →	**Irreversible Cell Injury**
Organ 'primed' for 'reperfusion injury' →	**Vascular Injury** - Poor Reflow - Odema

Fig. 7. Summary of some biochemical changes associated with ischaemia and reperfusion.

Conclusion

It is unlikely that a single biochemical event is responsible for the deterioration of organs subjected to prolonged periods of ischaemia. The investigations described in this chapter show that oxidative damage following renal ischaemia can be significantly inhibited by many different agents which suggests that a complex interaction of a number of factors are responsible for post-ischaemic tissue damage. Evidence has been presented for a number of intracellular changes which occur in renal tissue during ischaemia (summarised in Fig. 7). While all these changes may be considered adverse to normal cell function, they are nevertheless reversible within certain limits. Thus upon reperfusion, ATP may be regenerated and provide energy for calcium pumps and the restoration of calcium homeostasis. Similarly, delocalised intracellular iron also appears to be rapidly sequestered upon reoxygenation when the ischaemic period has been short. However, we believe these early events to be of crucial importance in 'priming' the organ for subsequent reperfusion damage and that the longer the period of ischaemia, the more significant these changes become.

Upon reperfusion, a burst of $O_2^{\cdot -}$ production from the incoming oxygen, due to a number of possible mechanisms already discussed, would react with increased levels of catalytic iron to yield much more reactive radical species. Damage to cellular components would ensue, including peroxidation of lipids in membranes already compromised by increased calcium-dependent phospholipase activity (Fig. 7). The resulting loss of integrity of the plasma membrane and intracellular organelles would cause further imbalances in intracellular ion homeostasis. This cycle of self-perpetuating events may perturb the cell sufficiently to cause irreversible cell injury.

In addition to 'gross' biochemical damage, effects on specific systems may contribute to post-ischaemic organ failure through disturbances in the vasculature of the organ (Fig. 7). These include imbalances in eicosanoid production due to a calcium-dependent build-up of free arachidonic acid, production of inflammatory mediators such as leukotrienes, release of chemotactic substances with subsequent adhesion and activation of polymorphonucleocytes, and derangement of receptor-mediated functions such as the phosphatidylinositol secondary messenger system.

The involvement of a wide spectrum of biochemical derangements in ischaemic/reperfusion injury is supported by the fact that many different pharmacological agents have been reported to afford at least some physiological protection to kidneys subjected to ischaemic insult. These include: iron-chelators (Paller *et al.*, 1988); free-radical scavengers, including superoxide dismutase (Paller *et al.*, 1984; Koyama *et al.*, 1985; Bosco & Schweizer, 1988) and catalase (Bosco & Schweizer, 1988); allopurinol (Koyama *et al.*, 1985); calcium-antagonists (Gingrich *et al.*, 1985; Schrier *et al.*, 1987); and prostacyclin analogues (Langkopf *et al.*, 1986). However, the wide range of biochemical abnormalities makes it unlikely that a single agent will be totally effective. We have investigated the effect of administering (i.v. both before ischaemia and before reperfusion) a combination of the iron-chelator DFX and the cyclooxygenase inhibitor indomethacin on the subsequent viability of kidneys rendered cold ischaemic (Gower *et al.*, 1989*a*). Kidneys were stored for up to 72 hr in HCA at 0 °C, autotransplanted into the contralateral renal bursa and the animals recovered. After 48 hr storage, 3/6 of the kidneys in untreated animals were capable of supporting life whereas all (6/6) of the treated group survived the full 30 days of the experimental period. Following the more damaging 72 hr period of CI, only 2/6 of the control group survived compared with 4/6 in the group treated with DFX and indomethacin. Analysis of serum urea and creatinine levels showed that all the failures died of uremia. While the groups were too small for these preliminary results to achieve significance, they did indicate a trend towards improved viability of kidneys subjected to clinically relevant periods of CI in a solution widely used for the storage of human organs.

Now that many of the immunological problems associated with organ transplantation can be successfully controlled, retrieval of donor organs in optimum condition is becoming an increasingly important aspect of clinical renal transplantation. Hypothermic storage in special solutions with improved ionic composition has already yielded benefits (Belzer & Southard, 1988). It is envisaged that further advances towards increasing both the number and post-storage viability of transplanted organs will come through a better understanding of the underlying causes of ischaemic/reperfusion damage. A combined pharmacological strategy to prevent, or at least slow, several adverse biochemical changes associated with ischaemia including altered intracellular iron and calcium homeostasis, loss of antioxidant protection and increased capacity for free-radical generation, would appear to offer great potential.

References

Aisen, P. & Listowsky, I. (1980). Iron transport and storage proteins. *Annual Reviews of Biochemistry* **49**, 357–93.

Arnold, P.E., Lumlertgul, D., Burke, T.J. & Schrier, R.W. (1985). In vitro versus in vivo mitochondrial calcium loading in ischemic acute renal failure. *American Journal of Physiology* **248**, F845–50.

Bakkeren, D.L., Jeu-Jaspars, C.M.H., Van der Heul, C. & Van Eijk, H.G. (1985). Analysis of iron-binding components in the low molecular weight fraction of rat reticulocyte cytosol. *International Journal of Biochemistry* **17**, 925–30.

Bellomo, G., Richelmi, P., Mirabelli, F., Marinoni, V. & Abbagnano, A. (1985). Inhibition of liver microsomal calcium ion sequestration by oxidative stress: role of protein sulphydryl groups. In *Free Radicals in Liver Injury*, ed. G. Poli, K.H. Cheeseman, M.U. Dianzani & T.F. Slater, pp. 139–42. Oxford: IRL Press.

Belzer, F.O. & Southard, J.H. (1988). Principles of solid-organ preservation by cold storage. *Transplantation* **45**, 673–6.

Berridge, M.J. (1984). Inositol triphosphate and diacylglycerol as second messengers. *Biochemical Journal* **220**, 345–60.

Berridge, M.J., Dawson, R.M.C., Downes, C.P., Heslop, J.P. & Irvine, R.F. (1983). Changes in the levels of inositol phosphates after agonist-dependent hydrolysis of membrane phosphoinositides. *Biochemical Journal* **212**, 473–82.

Biemond, P., Swaak, A.J.G., van Eijk, H.G. & Koster, J.F. (1988). Superoxide-dependent iron release from ferritin in inflammatory diseases. *Free Radical Biology and Medicine* **4**, 185–98.

Bolli, R. (1988). Oxygen-derived free radicals and postischemic myocardial dysfunction ('stunned myocardium'). *Journal of the American College of Cardiology* **12**, 239–49.

Bosco, P.J. & Schweizer, R.T. (1988). Use of oxygen radical scavengers on autografted pig kidneys after warm ischemia and 48-hour perfusion preservation. *Archives of Surgery* **123**, 601–4.

Calman, K.C., Quin, R.O. & Bell, P.R. (1973). Metabolic aspects of organ storage and the prediction of organ viability. In *Organ Preservation*, ed. D.E. Pegg, pp. 225–40. London: Churchill Press.

Carafoli, E. (1987). Intracellular calcium homeostasis. *Annual Reviews of Biochemistry* **56**, 395–433.

Cheung, J.Y., Bonventre, J.V., Malis, C.D. & Leaf, A. (1986). Calcium and ischemic injury. *New England Journal of Medicine* **314**, 1670–6.

Cooper, C.L. & Malik, K.U. (1985). Prostaglandin synthesis and renal vasoconstriction elicited by adrenergic stimuli are linked to activation of alpha-1 adrenergic receptors in the isolated rat kidney. *Journal of Pharmacology and Experimental Therapeutics* **233**, 24–31.

Cotterill, L.A., Gower, J.D., Fuller, B.J. & Green, C.J. (1989*a*). Oxidative damage to kidney membranes during cold ischaemia: Evidence of a role for calcium. *Transplantation* **48**, 745–51.

Cotterill, L.A., Gower, J.D., Fuller, B.J. & Green, C.J. (1989*b*). Oxidative stress during hypothermic storage of rabbit kidneys: possible mechanisms by which calcium mediates free radical damage. *Cryo-Letters* **10**, 119–26.

Cotterill, L.A., Gower, J.D., Fuller, B.J. & Green, C.J. (1989*c*). Free fatty acid accumulation following cold ischaemia in rabbit kidneys and the involvement of a calcium dependent phospholipase A_2. *Cryo-Letters* **11**, 3–12.

Demopoulos, H.B., Flam, E.S., Pietronigro, D.D. & Seligman, M. (1980). The free radical pathology and the micro-circulation in the major central nervous system disorders. *Acta Physiologica Scandinavica* **492**, 91–119.

Dianzani, M.U., Paradisi, L., Barrera, G., Rossi, M.A. & Parola, M. (1989). The action of 4-hydroxynonenal on the plasma membrane enzymes from rat hepatocytes. In *Free Radicals, Metal Ions and Biopolymers*, ed. P.C. Beaumont, D.J. Deeble, B.J. Parsons & C. Rice-Evans, pp. 329–46. London: Richelieu Press.

Dunford H.B. (1987). Free radicals in iron-containing systems. *Free Radical Biology and Medicine* **3**, 405–21.

Evered, D. & Whelan, J. (eds.) (1986). *Calcium and the Cell, Ciba Foundation Symposium 122*. Chichester: John Wiley & Sons.

Farber, J.L. (1981). The role of calcium in cell death. *Life Sciences* **29**, 1289–95.

Fuller, B.J., Gower, J.D. & Green, C.J. (1988). Free radicals and organ preservation: fact or fiction? *Cryobiology* **25**, 377–93.

Funk, F., Lenders, J.-P., Crichton, R.R. & Schneider, W. (1985). Reductive mobilisation of ferritin iron. *European Journal of Biochemistry* **152**, 167–72.

Gingrich, G.A., Barker, G.R., Lui, P. & Stewart, S.C. (1985). Renal preservation following severe ischemia and prophylactic calcium channel blockade. *Journal of Urology* **134**, 408–10.

Goddard, J.G., Serebin, S., Basford, D. & Sweeney, G.D. (1986). Microsomal chemiluminescence (lipid peroxidation) is stimulated by oxygenation in the presence of anaerobically released ferritin iron. *Federation Proceedings* **45**, 1746.

Gower, J.D., Healing, G., Fuller, B.J., Simpkin, S. & Green, C.J. (1989*a*). Protection against oxidative damage in cold-stored rabbit kidneys by desferrioxamine and indomethacin. *Cryobiology* **26**, 309–17.

Gower, J.D., Healing, G. & Green, C.J. (1989*b*). Determination of desferrioxamine-available iron in biological tissues by high-pressure liquid chromatography. *Analytical Biochemistry* **180**, 126–30.

Gower, J., Healing, G. & Green, C. (1989*c*). Measurement by HPLC of desferrioxamine-available iron in rabbit kidneys to assess the effect of ischaemia on the distribution of iron within the total pool. *Free Radical Research Communications* **5**, 291–9.

Granger, D.N., Benoit, J.N., Suzuki, M. & Grisham, M.B. (1989). Leukocyte adherence to venular endothelium during ischaemia–reperfusion. *American Journal of Physiology* G683–8.

Green, C.J., Healing, G., Lunec, J., Fuller, B.J. & Simpkin, S. (1986*a*). Evidence of free radical-induced damage in rabbit kidneys after simple hypothermic preservation and autotransplantation. *Transplantation* **41**, 161–5.

Green, C.J., Healing, G., Simpkin, S., Fuller, B.J. & Lunec, J. (1986*b*). Reduced susceptibility to lipid peroxidation in cold ischaemic rabbit kidneys after addition of desferrioxamine, mannitol or uric acid to the flush solution. *Cryobiology*, **23**, 358–65.

Green, C.J., Healing, G., Simpkin, S., Lunec, J. & Fuller, B.J. (1986*c*). Desferrioxamine reduces susceptibility to lipid peroxidation in rabbit kidneys subjected to warm ischaemia and reperfusion. *Comparative Biochemistry and Physiology* **85**B, 113–17.

Gutteridge, J.M.C. (1984). Lipid peroxidation initiated by superoxide-dependent hydroxyl radicals using complexed iron and hydrogen peroxide. *FEBS Letters* **172**, 245–9.

Gutteridge, J.M.C., Richmond, R. & Halliwell, B. (1979). Inhibition of the iron-catalysed formation of hydroxyl radicals from superoxide and of lipid peroxidation by desferrioxamine. *Biochemical Journal* **184**, 469–72.

Halliwell, B. (1978). Superoxide-dependent formation of hydroxyl radicals in the presence of iron salts. *FEBS Letters* **96**, 238–42.

Halliwell, B. & Gutteridge, J.M.C. (1984). Oxygen toxicity, oxygen radicals, transition metals and disease. *Biochemical Journal* **219**, 1–14.

Healing, G., Gower, J.D., Fuller, B.J. & Green, C.J. (1990). Intracellular iron redistribution: an important determinant of reperfusion damage to rabbit kidneys. *Biochemical Pharmacology* **39**, 1239–45.

Hesse, I.F.A. & Johns, E.J. (1984). The subtype of α-adrenoceptor involved in the neural control of renal tubular sodium reabsorption in the rabbit. *Journal of Physiology* **328**, 527–38.

Irvine, R.F., Anggard, E.E., Letcher, A.J. & Downes, C.P. (1985). Metabolism of inositol 1,4,5-trisphosphate and inositol 1,3,4-trisphosphate in rat parotid glands. *Biochemical Journal* **229**, 505–11.

Isakson, P.C., Raz, A., Hsueh, W. & Needleman, P. (1978). Lipases and prostaglandin biosynthesis. In *Advances in Prostaglandin and Thromboxane Research*, vol. 3, ed. C. Galli, pp. 113–19. New York: Raven Press.

Jewell, S.A., Bellomo, G., Thor, H., Orrenius, S. & Smith, M.T. (1982). Bleb formation in hepatocytes during drug metabolism is caused by disturbances in thiol and calcium ion homeostasis. *Science* **217**, 1257–9.

Keberle, H. (1964). The biochemistry of desferrioxamine and its relation to iron metabolism. *Annals of the New York Academy of Sciences USA* **119**, 758–68.

Kessar, P. & Saggerson, E.D. (1980). Evidence that catecholamines stimulate renal gluconeogenesis through an α_1-type of adrenoceptor. *Biochemical Journal* **190**, 119–23.

Koyama, I., Bulkley, G.B., Williams, G.M. & Im, M.J. (1985). The role of oxygen free radicals in mediating the reperfusion injury of cold-preserved ischemic kidneys. *Transplantation* **40**, 590–5.

Langkopf, B., Rebmann, U., Schabel, J., Pauer, H.-D., Heynemann, H. & Forster, W. (1986). Improvement in the preservation of ischemically impaired renal transplants of pigs by iloprost (ZK 36, 374). *Prostaglandins, Leukotrienes and Medicine* **21**, 23–8.

Laurent, B. & Ardaillou, R. (1986). Reactive oxygen species: production and role in the kidney. *American Journal of Physiology* **251**, F765–76.

Lefer, A.M. (1985). Eicosanoids as mediators of ischaemia and shock. *Federation Proceedings* **44**, 275–80.

Lelcuk, S., Alexander, F., Kobzik, L., Valeri, C.R., Shepro, D. & Hechtman, H.B. (1985). Prostaglandins and thromboxane A2 moderate post-ischaemic renal failure. *Surgery* **98**, 207–12.

Matsumura, Y., Miyawaki, N., Sasaki, Y. & Morimoto, S. (1985). Inhibitory effects of norepinephrine, methoxamine and phenylephrine on renin release from rat kidney cortical slices. *Journal of Pharmacology and Experimental Therapeutics* **233**, 782–7.

McCord, J.M. (1985). Oxygen-derived free radicals in post-ischaemic tissue injury. *New England Journal of Medicine* **312**, 159–63.

Minotti, G. & Aust, S.D. (1987). The role of iron in the initiation of lipid peroxidation. *Chemistry and Physics of Lipids* **44**, 191–208.

Momchilova, A., Petkova, D. & Koumanov, K. (1986). Rat liver microsomal phospholipase A2 and membrane fluidity. *International Journal of Biochemistry* **18**, 659–63.

Monteiro, H.P. & Winterbourn, C.C. (1988). The superoxide-dependent transfer of iron from ferritin to transferrin and lactoferrin. *Biochemical Journal* **256**, 923–8.

Mulligan, M., Althaus, B. & Linder, M.C. (1986). Non-ferritin, non-heme iron pools in rat tissues. *International Journal of Biochemistry* **18**, 791–8.

Nayler, W.G., Panagiotopoulos, S., Elz, J.S. & Daly, M.J. (1988). Calcium-mediated damage during post-ischaemic reperfusion. *Journal of Molecular and Cellular Cardiology* **20**, Suppl. 2, 41–54.

Nicotera, P.L., Moore, M., Mirabelli, F., Bellomo, G. & Orrenius, S. (1985). Inhibition of hepatocyte plasma membrane Ca^{2+} ATPase activity by menadione metabolism and its restoration by thiols. *FEBS Letters* **181**, 149–53.

Opie, L.H. (1989). Proposed role of calcium in reperfusion injury. *International Journal of Cardiology* **23**, 159–64.

Paller, M.S., Hoidal, J.R. & Ferris, T.F. (1984). Oxygen free radicals in ischaemic acute renal failure in the rat. *Journal of Clinical Investigation* **74**, 1156–64.

Paller, M.S., Hedlund, B.E., Sikora, J.J., Faassen, A. & Waterfield, R. (1988). Role of iron in postischemic renal injury in the rat. *Kidney International* **34**, 474–80.

Pavlock, G.S., Southard, J.H, Starling, J.R. & Belzer, F.O. (1984). Lysosomal enzyme release in hypothermically perfused dog kidneys. *Cryobiology* **21**, 521–8.

Ratych, R.E. & Bulkley, G.B. (1986). Free-radical-mediated postischemic reperfusion injury in the kidney. *Journal of Free Radicals in Biology and Medicine* **2**, 311–19.

Robak, J. & Sobanska, B. (1976). Relationship between lipid peroxidation and prostaglandin generation in rabbit tissues. *Biochemical Pharmacology* **25**, 2233–6.

Ross, H., Marshall, V.C. & Escott, M.L. (1976). 72-hr canine kidney preservation without continuous perfusion. *Transplantation* **21**, 498–501.

Roy, R.S. & McCord, J.M. (1983). Superoxide and ischaemia: conversion of xanthine dehydrogenase to xanthine oxidase. In *Oxyradicals and Their Scavenging Systems*, vol. 2, ed. R. Greenwald & G. Cohen, pp. 145–53. New York: Elsevier Science.

Schieppati, A., Wilson, P.D., Burke, T.J. & Schrier, R.W. (1985). Effect of renal ischaemia on cortical microsomal calcium accumulation. *American Journal of Physiology* **249**, C476–83.

Schlondorff, D. & Ardaillon, R. (1986). Prostaglandins and other arachidonic acid metabolites in the kidney. *Kidney International* **29**, 108–19.

Schmitz, J.M., Apprill, P.G., Buja, L.M., Willerson, J.T. & Campell, W.B. (1985). Vascular prostaglandin and thromboxane production in a canine model of myocardial ischaemia. *Circulation Research* **57**, 223–31.

Schoenberg, M., Younes, M., Muhl, E., Sellin, D., Fredholm, B. &

Schildberg, F.W. (1983). Free radical involvement in ischaemic damage to the small intestine. In *Oxyradicals and Their Scavenger Systems*, vol. 2, ed. R. Greenwald & G. Cohen, pp. 154–7. New York: Elsevier Science.

Schrier, R.W., Arnold, P.E., van Putten, V.J. & Burke, T.J. (1987). Cellular calcium in ischemic acute renal failure: role of calcium entry blockers. *Kidney International* **32**, 313–21.

Sehr, P.A., Bore, P.J., Papatheofanis, J. & Radda, G.K. (1979). Non-destructive measurement of metabolites and tissue pH in the kidney by ^{31}P nuclear magnetic resonance. *British Journal of Experimental Pathology* **60**, 632–41.

Southard, J.H., Ametani, M.S., Lutz, M.F. & Belzer, F.O. (1984). Effects of hypothermic perfusion of kidneys on tissue and mitochondrial phospholipids. *Cryobiology* **21**, 20–4.

Ungemach, F.R. (1985). Plasma membrane damage of hepatocytes following lipid peroxidation: involvement of phospholipase A2. In *Free Radicals in Liver Injury*, ed. G. Poli, K.H. Cheeseman, M.U. Dianzani & T.F. Slater, pp. 127–34. Oxford: IRL Press.

Vile, G.F. & Winterbourn, C.C. (1988). Adriamycin-dependent peroxidation of rat liver and heart microsomes catalysed by iron chelates and ferritin. *Biochemical Pharmacology* **37**, 2893–7.

Weltzem, H.U. (1979). Cytolytic and membrane-perturbing properties of lysophosphatidylcholine. *Biochimica et Biophysica Acta* **559**, 259–87.

Wills, E.D. (1969). Lipid peroxide formation in microsomes: the role of non-heam iron. *Biochemical Journal* **113**, 325–32.

ANTHONY K. CAMPBELL

The Rubicon Hypothesis: a quantal framework for understanding the molecular pathway of cell activation and injury

Prologue

Calcium and oxygen are two of the major elements in the earth's crust (Campbell, 1983). Both are essential for maintaining the earth's ecosystem, and in particular for the existence and function of the human body. A further common link between calcium and oxygen in living systems is found in the biological molecules which bind Ca^{2+}, where usually oxygen is the atom coordinating the Ca^{2+} (Duncan, 1976; Campbell, 1983). Yet both O_2 and Ca^{2+} can be toxic, and even lethal, to cells and whole organisms. In spite of the abundance of calcium and oxygen in the substances which make up the earth's crust it is likely that, when life began 3500 million years ago, the concentration of free Ca^{2+} and O_2 surrounding the first cells was much lower than it is today. Throughout evolution the physiology of cells has been intimately linked to their pathology, because of the necessity of developing defence mechanisms against chemical attack from Ca^{2+} or oxygen radicals, as well as other damaging chemical, physical and biological agents.

It has been known for nearly a century that a rise in cytosolic free Ca^{2+} is responsible for initiating cellular events such as movement, secretion, transformation and division (Heilbrunn, 1937; 1956; Duncan, 1976; Campbell, 1983, 1989*a*; Carafoli, 1987; Rasmussen, 1989; Reid, Cook & Luzio, 1989), and for certain defence mechanisms such as vesiculation (Campbell & Morgan, 1985; Morgan, 1989). Yet a prolonged, high level of intracellular free Ca^{2+} can irreversibly damage mitochondria, can cause chromatin condensation, precipitation of phosphate and protein and activate degradative enzymes such as proteases, nucleases and phospholipases. Similarly an increase in oxygen supply is necessary for cells whose energy demands increase, for example during muscle contraction. Yet culturing some cells in pure O_2 and even sometimes in air (20% O_2), is lethal, a fact discovered by Priestley some 200 years ago.

How then are we to tell whether in a particular circumstance Ca^{2+} or

O_2 is friend or foe (Campbell, 1987; Halliwell & Gutteridge, 1989)? The reductionist approach is to search diligently for ever more fine molecular detail within cells and tissues attacked by intracellular Ca^{2+} or by oxygen and its metabolites. There is no doubt that this has been highly successful at identifying and characterising the molecules, and the reactions, required to keep a cell alive, and to enable the cell to carry out its specialised functions. In this article I attempt to take a more holistic stance. Perhaps the molecular cell biologist's equivalent to the global Gaia hypothesis (Lovelock, 1988; Baerlocher, 1990). The essential feature of this approach is that the unit of life is the cell, not the biochemist's homogenate or even DNA. Unlike the homogenate, where reactions appear continuous and change in a smooth or graded manner, in the live cell the processes essential for life exhibit discontinuities, and behave in a quantal manner. Only when we can step back and view all the small pieces of the jigsaw completing one picture, all the small quantum leaps which lead to a threshold end response, will we have our first real glimpse of the answer to the central question of biology. What is unique about the chemistry and physics of a cell which determines whether we assign it as live or dead?

The problem

At the turn of the nineteenth century, human understanding of the physical nature of the universe was revolutionised by the concept, introduced by Planck and developed by Einstein, that energy is not continuous. Rather, energy exists in discrete packets, or quanta. The search for this new idea had been provoked by a 'crisis point' in physics, the so-called 'violet catastrophe', a conflict between the description by the experimentalist of the spectrum of light emitted from a hot body and that predicted mathematically by the then theory. Crisis points in science force creative minds to think laterally. Mitchell (1966) found chemisymbiosis a radical solution to the 'crisis' provoked by the failure of conventional biochemists to find the expected chemical intermediate in mitochondria, linking substrate oxidation to ATP synthesis. He derived a simple formula, in this case relating 'energy potential' for ATP synthesis to membrane potential and a pH gradient.

A universal property of eukaryotic cells, and under some circumstances prokaryotes, is their ability to change state as a result of a physical or chemical event initiated at the plasma membrane (Table 1). The initial event at the cell surface includes touch, the arrival of an action potential, the binding of a hormone or neurotransmitter to receptors, the binding of an antibody to an antigen and the attachment of a T cell to its antigenic

Table 1. *Examples of external cell signals*

Signal	Cell stimulated	Response
Physical		
action potential	nerve terminal	neuro-secretion
touch	Paramecium	reversal of ciliate movement
light	photoreceptor	communication to CNS
Chemical		
blood clotting	platelet	aggregation
chemoattractant	neutrophil	chemotaxis
neurotransmitter	skeletal muscle	contraction
hormone	hepatocyte	glucose release
substrate glucose	B cell	insulin secretion
paracrine	smooth muscle	contraction
cytokine	lymphocyte	cell division and antibody production
antibody + complement	erythrocyte	lysis
oxygen metabolite	bacterium	death
Biological		
cell–cell adherence (e.g. killer T cell)	tumour cell	death by apoptosis
bacterium	neutrophil	phagocytosis
virus	endothelial cell	transformation
trypanosome	erthrocyte	uptake

target, and the insertion into the membrane of pore-forming proteins from the complement system, bacterial toxins or viral proteins. End responses evoked by these agents include cell movement, secretion, phago- and endo-cytosis, aggregation, transformation, division, activation of metabolism, specialised phenomena such as the generation of O_2^- by phagocytes and bioluminescence, cell defence and cell death. Much has been learnt about the molecular pathway linking the initial interaction with the plasma membrane and the ultimate end response (Fig. 1). Firstly a chemical signal or signals are generated in the plasma membrane or released from the inner surface. The first such signals to be discovered were Ca^{2+} and H^+ (Heilbrunn, 1937; 1956; see Campbell 1983; 1987; 1988; 1989*a*; Reid *et al.*, 1989 for references). The last thirty years has seen the discovery of many more, including cyclic nucleotides, inositol phosphates, and other phospholipid derivatives such as diacyl glycerol (Berridge & Irvine 1984; 1990), as well as oscillations in cytosolic Ca^{2+}

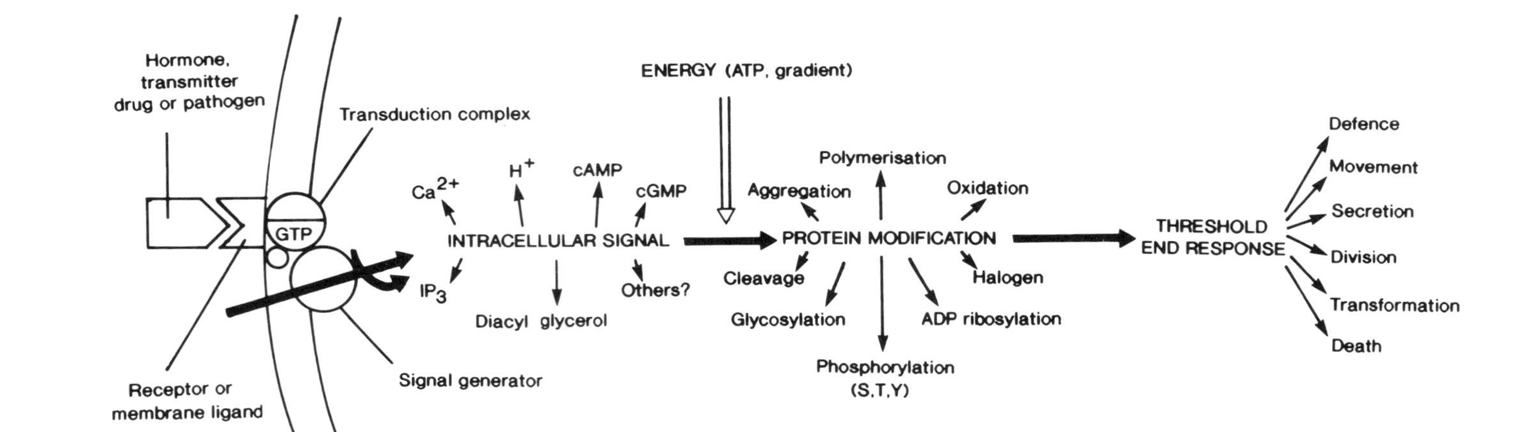

Fig. 1. From plasma membrane to end response.

(Woods, Cuthbertson & Cobbold, 1986; Berridge & Galione, 1988; Kanno, Saito & Yamashita, 1989). These signals bind to intracellular molecules leading to reorganisation or covalent modification of proteins, other macromolecules and organelles which are responsible for producing the cell's response. These modifications include phosphorylation of ser, thr, and tyr residues, ADP ribosylation, proteolytic cleavage, transglutamination, and halogenation from radical reactions. We know a lot about many of the proteins and the enzymes catalysing reactions in these sequences; several have been cloned and sequenced: adenylate and guanylate cyclase, the transducer or G proteins (Gilman, 1987), kinases and their recognition peptides (Cohen, 1988), phospholipases, Ca^{2+}-binding proteins such as calmodulin and troponin C with their EF-hand high affinity, selective Ca^{2+} sites (Persechini, Moncrief & Kretsinger, 1989). Three-dimensional (3D) structural analysis, combined with protein engineering, has given us some idea of how these proteins do their job. Genes such as the homeobox family and oncogenes, and their protein products, which play a key role in the development of organism structure and in its breakdown in cancer have been characterised (Akam, 1987). These are all part of the signal-transduction system which controls the behaviour and structure of individual eukaryotic cells and their groupings into organs.

Yet in spite of the undoubted success of this reductionist approach the problems at the heart of the phenomenon being studied remain. In no system do we have a complete molecular sequence from the initial interaction of the cell with an agonist, drug or pathogen, to end response. Many eukaryotic cells have the ability to produce different types of response depending on the stimulus. What determines whether and when a cell undergoes a particular response? For example, the same primary external stimuli can provoke the neutrophil to move, aggregate, secrete proteins, produce O_2^- and other oxygen metabolites or phagocytose. Yet every stimulus seems to provoke release of the usual intracellular signals; IP_3, Ca^{2+}, cyclic AMP, diacyl glycerol. The molecular basis of the diseases which are the major afflictions of western society, namely atheroma, arthritis and other inflammatory or immune-based disease, neurodegenerative diseases such as multiple sclerosis and motor neurone disease, cancer, inherited disorders such as cystic fibrosis and muscular dystrophy, remain elusive. This, in spite of the intensity of effort and discovery of important details such as the identification of key phosphorylated proteins or the site of mutation. Furthermore with specific reference to the subject of this book, in spite of volumes of publications describing changes in biological molecules and structures produced by oxygen radicals or Ca^{2+}, the significance of many of these to a real physiological or pathological sequence remains unestablished.

Is this then a 'crisis point' or are the complete molecular solutions just around the reductionists' corner? Is the increasing documentation of more and more intracellular signals and a plethora of phosphorylated proteins during a cell response clarifying or confusing the problem? Should we be focussing on one key event, one particular signal, one particular modified protein? The heart of the matter is that we are not viewing the phenomena, nor the molecular changes associated with them, in a way which will enable us to reconstruct a complete molecular sequence from plasma membrane to end response. There are two reasons for this, one conceptual, the other experimental. Firstly the pathway is not a smooth one, and secondly the timing of the discontinuities which make up the essential components of the molecular sequence vary from cell to cell. Thus biochemical measurements on populations of cells, which have evolved to behave asynchronously, will inevitably smooth out these vital discontinuities and they will be missed. Furthermore, without temporal single cell analysis, it is impossible to ensure that an intracellular chemical change, such as Ca^{2+} or oxidation of protein or lipid by an oxygen metabolite, is a cause or consequence of the ultimate cell response.

It is therefore clear that a better conceptual framework is needed in order to link the extracellular signals with the network of intracellular signals and protein modifications generated by them, and then fit them to the end response. This must be based on identifying and examining the essential units responsible for the event, i.e. the molecular groupings, the organelle, and the unit of life, the individual cell itself. Furthermore, techniques are required to measure and locate chemical changes in individual cells.

The Rubicon Hypothesis

A solution?

The central feature of this hypothesis is that the cell is the unit of life, and that each structure within it, e.g. an organelle such as mitochondrion, must be considered as an individual thermodynamic entity. The maintenance of a live cell, the control of its behaviour, its development, its reproduction, and its eventual death, are determined by a series of physical and chemical thresholds, or 'rubicons'. Only when a specific series of rubicons is crossed at the right time, in the correct order, and at the right location within the cell, will a particular cellular event occur. Similarly the behaviour of an organelle is also determined by the 'rubicons' it has crossed. This means that in a population of cells, activated by an agonist or drug, or attacked by a pathogen, at any one time some cells will have crossed all the rubicons necessary for the end response, some but a few on the way to an end response, and some will have crossed none. Even within one cell, in a group of organelles or sub-cellular structures, where the number of rubicons to be crossed is less than for the whole cell, there will be some which have crossed all their rubicons whilst some will still be in the state they were before the cell was attacked. The sequence is initiated by an event at the plasma membrane as the result of exposure to an external agent, or by an event within the cell resulting from internal programming. Between each 'rubicon' there is a physio-chemical pathway to the banks of the next rubicon. Then, another set of physical and chemical changes takes the cell or organelle across. The pathway to the bank of the next rubicon then begins. In some cases the crossing will be much faster than the time taken to reach the crossing point, particularly in acute regulatory events such as oxygen-radical generation, movement or secretion. However, this is not always so, for in more longer term regulation, e.g. cellular development and differentiation, the crossing may be the longest step in the sequence.

An important consequence of this hypothesis is that it is invalid to make chemical and physical measurements on populations of cells or organelles, or to derive mathematical relationships from these data, if the complete molecular basis of cell activation or injury is to be elucidated. Thus the job of the molecular cell biologist is to unravel the chemical and physical basis of the route between each rubicon, the molecular basis of each crossing, and what is the route of return.

Is this hypothesis really saying anything new or is it simply a restatement of the obvious? Does it genuinely give us new perspective on the chemistry and physics of cell activation and injury, and on life itself? The answers to these questions lie in the answers to six specific questions.

1. Is there any evidence to support the hypothesis? In particular, have any 'rubicons' already been identified?
2. What predictions does the hypothesis make, which will enable us not only to benefit but also to make important new discoveries?
3. What is the experimental strategy required to test these predictions, and is there any new methodology required?
4. Does the hypothesis really help us to rationalise, and understand better, the molecular and cellular biology of intracellular signalling, and in particular the role of Ca^{2+} and oxygen radicals in cell activation and injury?
5. What are the consequences for understanding the molecular basis of disease?

6. In this quantal hypothesis, is it possible to make the step from qualitative to quantitative, and derive a mathematical relationship between particular chemical and physical components in a cell and the timing and magnitude of each threshold, and the ultimate end response?

Let us therefore examine these questions, and in particular the relevence to Ca^{2+} and oxygen radicals: the theme of this book.

The evidence

In the search for evidence we must first examine the nature of end responses in cells (Table 2), and then the nature of the signals and organelle responses (Table 3) responsible for them. To state that a cell which has undergone an end response must have crossed at least one threshold is stating the obvious (Table 2). A muscle cell contracts or remains relaxed, a heart cell beats, a platelet aggregates with a neighbour during blood clotting or remains free, an excitable cell generates an action potential or it does not, a luminous cell flashes or remains invisible, an endocrine cell secretes or its hormone remains within an intracellular vesicle, one cell becomes two following division, a cell dies or recovers from injury. What is different about the Rubicon Hypothesis is that it tells us something new about the timing and magnitude of such end responses. The graphic representation of a time course or dose response appears smooth, but in reality involves cells crossing the end response at different times, rather than there being a gradual, smooth increase in the magnitude of the response. At half maximum have all the cells been activated to 50% of their maximum or have only half the cells crossed the end response rubicon?

Small potential changes can be detected at the end plate of a muscle fibre, because of the spontaneous release of ACh from vesicles in the nerve terminal. Only when the rate of occupancy of receptors is fast enough, following the burst of release from the nerve action potential, is an action potential generated in the muscle, which then releases its internal Ca^{2+} store and contracts. In the nerve cell body stimulatory and inhibitory transmitters synergise or antagonise each other. Only when the net sum of the potential is sufficient will an action potential be generated and travel down the axon.

In a population of neutrophils activated by chemotactic or phagocytic stimuli there is a burst of O_2 uptake some 10–30 s following addition of the stimulus. This slows after 30–60 s but remains at a level of O_2 uptake higher than the resting cells for many minutes, as long as O_2 is still available. Measurement of the $O_2^{\cdot-}$ released from the activated oxidase

Table 2. *Some phenomena described by a threshold end response in individual cells*

Threshold in the individual cell	Cellular example
1. Movement	
muscle contraction	skeletal muscle
chemotaxis	neutrophil
phototaxis	bacterium
2. Electrical activity	
action potential generation	neurone cell body
3. Adherence	
blood clotting	platelet–platelet
antibody production	T cell–B cell
infection	neutrophil–bacterium
4. Release from cells	
secretion from vesicles	nerve terminal
nematocyst discharge	jelly–fish cnidocyte
virus release	epithelial cell
release of $O_2^{\overline{\cdot}}$	neutrophil
light in bioluminescence	luminous dinoflagellate flash
5. Transformation	
antibody production	B cell→plasma cell
Nigleria	amoeboid→flagellate
cancer	lymphocytic leukaemia
fertilisation	sperm→egg + sperm
oocyte maturation	starfish oocyte
cell division	many cells
6. Cell injury	
vesiculation and blebbing	myocyte
reversible cell damage by pore formers	neutrophil
lysis by pore formers	erthyrocyte
demyelination	oligodendrocyte
death by apoptosis	T lymphocyte

responsible for this burst, using the bioluminescent indicator pholasin (Roberts, Knight & Campbell, 1987), or measurement of intracellular H_2O_2 formation using the fluor 2,7-dichlorofluorescin (Patel, Hallett & Campbell, 1987*a*), shows that the time course is really described by

Table 3. *Some examples of thresholds in intracellular organelles*

Organelle	Threshold response
nucleus	chromosome appearance chromosome movement loss of nuclear membrane gene expression condensation of chromatin in injury or death
endoplasmic reticulum (ER)	release of Ca^{2+} into cytosol
ribosome + ER	binding to mRNA
ribosome	release of completed protein
cytoplasmic vesicles	fusion with plasma membrane fusion with another vesicle
mitochondria	condensation division
cytoskeleton	contraction and movement
cytosol	protoplasmic streaming
gap junction	on/off of cell communication
plasma membrane	capping or clustering of proteins formation of endocytotic vesicle formation of phagosome fusion with secretory vesicle release of specialised organelles opening of ion channels permeabilisation without lysis lysis

individual cells switching on the oxidase at different times (Fig. 2). Furthermore, the decrease in O_2^- production caused by adenosine, which only inhibits stimuli dependent on a rise in cytosolic Ca^{2+} (Roberts *et al.*, 1985*a*), is explained mainly by a reduction in the number of activated cells, rather than a reduction in the magnitude of oxidase activity in each cell (Fig. 2). Similarly the phorbol ester, phorbol myristate acetate (PMA), also acts by switching on the oxidase at different times, but the number of cells activated during the first few minutes of exposure to this stimulus is much less than with the chemotactic peptide N formyl met leu

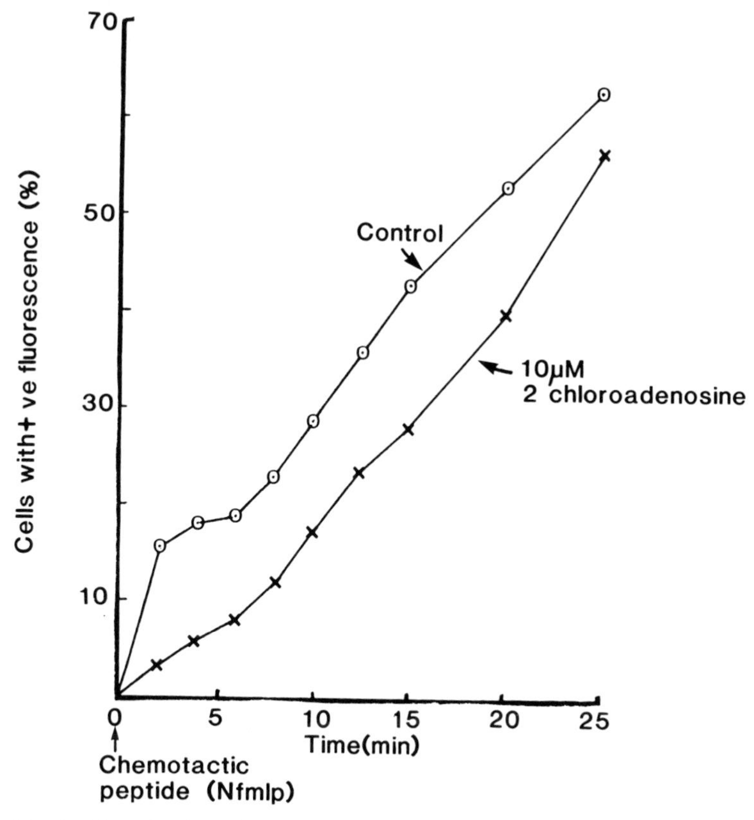

Fig. 2. Threshold for $O_2^{\overline{\cdot}}$ in neutrophils. Human neutrophils containing the fluorescent indicator 2′7′-dichlorofluorescin were incubated with the chemotactic peptide Nfmlp (1 μM) ± 10 μM 2 chloroadenosine. The number of activated cells, i.e. those with positive fluorescence, at times up to 30 min. were measured using the fluorescence activated cell sorter. From Patel *et al.* (1987*a*); copyright The Biochemical Society, with permission.

phe (Nfmlp). Thus the time course of $O_2^{\overline{\cdot}}$ production measured in a population of cells appears much slower with PMA than with Nfmlp. But what of the intracellular signals which initiate the reactions within the neutrophil leading to $O_2^{\overline{\cdot}}$ production?

Measurement of cytosolic free Ca^{2+} using the photoprotein obelin in populations of neutrophils identified two groups of stimuli (Hallett & Campell, 1982; Campbell & Hallett, 1983; Campbell *et al.*, 1988). Chemotactic peptides such as Nfmlp or C5a cause a rise in cytosolic free Ca^{2+}, which is necessary for oxidase activation, whereas phagocytic stimuli such as uncoated particles or phorbol esters cause no rise in cytosolic free Ca^{2+}. These results have been confirmed using fluorescent Ca^{2+} indicators such as fura 2 (Lew *et al.*, 1987), though some phagocytic stimuli may cause a small rise in cytosolic Ca^{2+} (Krustal, Shak & Maxfield, 1986; Krustal & Maxfield, 1987; Marks & Maxfield, 1990). Single cell fluorescent imaging has confirmed that the main source of the Ca^{2+} rise for Nflmp is extracellular, via an increase in permeability to the plasma membrane. An important characteristic is that there is a threshold for the Ca^{2+} rise, when observed in each cell, supporting the Rubicon Hypothesis. Depending on the cell, it takes 6–56 s for the cytosolic Ca^{2+} concentration to begin to increase (Hallett, Davies & Campbell, 1990). In the majority, but not all, of the cells, activation of the oxidase is tightly coupled to Ca^{2+}, there being a threshold Ca^{2+} concentration necessary to evoke $O_2^{\bar{\cdot}}$ release. A significant proportion of cells produce no detectable $O_2^{\bar{\cdot}}$, even at maximum Nfmlp. In these cells either no Ca^{2+} rise occurs, i.e. they fail to cross the Ca^{2+} rubicons, or the Ca^{2+} rise does not cross a critical threshold level. Another example of crossing the rubicon for neutrophils is seen when examining the release of internally stored Ca^{2+} in single cells, presumably caused by IP_3 (Davies, Hallett & Campbell, 1991). Intracellular Ca^{2+} is released from a single, highly localised internal store, not compatible with the cellular distribution of 'calciosomes' (Volpe *et al.*, 1988), in less than a third of the cells, even at maximum Nfmlp concentrations. This is less than the number of cells where cytosolic Ca^{2+} increases as a result of movement through the plasma membrane. The question now arises as to what the relevence of these Ca^{2+} rubicons, and those of other intracellular signals, are for the thresholds for primary and secondary granule secretion or for formation of a complete phagosome.

An even clearer example of the Rubicon Hypothesis in action is seen when examining the effect of the membrane attack complex of complement on erythrocytes, neutrophils and oligodendrocytes (Patel & Campbell, 1987). Complement is a system in the blood consisting of more than twenty proteins which plays a vital role in the body's defence against infection. The pathway is activated by antibody binding to surface antigens, or by an 'alternative' route, not requiring antibody. The terminal part of the pathway involves proteolytic cleavage of C5 to C5a and C5b. C5b binds C6 which inserts into the membrane of the cell being attacked. This is followed by C7, C8 and then up to 12–18 C9 molecules. The complete membrane attack complex, i.e. $C5b789_n$, then tries to lyse the cell. A typical time course for the release of haemoglobin from erythrocytes attacked by complement shows that more than 90% of the haemoglobin is released within 30–60 min. Examination of the cells under the electron microscope will reveal complexes on the surface with a 'pore' in the centre (Muller-Eberhard, 1988) just right, apparently for letting substances leak gradually out of the cell. There are two vital misconceptions associated with this model. Firstly, the complex seen on the surface, which contains a polymerised version of C9, is the inactive form (Dankert, Shiver & Esser, 1985; Dankert & Esser, 1987; Patel, Morgan & Campbell 1987*b*). Large molecules like proteins do not leak out through the membrane attack complex itself, but rather as a result of a major disruption in membrane permeability throughout the plasma membrane. Secondly, when we examine individual cells under the light microscope or with a cell sorter we see that the haemoglobin is not leaking out gradually from all the cells. Rather each cell crosses a critical threshold, a lytic rubicon, when the cell suddenly becomes highly permeable and releases all of its internal constituents very rapidly (Edwards *et al.*, 1983). The time at which a cell crosses the lytic rubicon can vary from a few minutes to half an hour or more. A crucial question, therefore, is what is the molecular reason for these differences in timing from cell to cell? All the cells appear to have membrane attack complexes on their surface within a few seconds, as detected by a fluorescein-labelled monoclonal antibody to C9.

It has been known for some years that cells from the same species as the complement source (autologous) are much more difficult to lyse than cells attacked by complement from another species. Furthermore, nucleated cells are much more difficult to lyse than aged erythrocytes, the classic system for studying complement membrane attack (Goldberg & Green, 1959; Green, Barrow & Goldberg, 1959; Boyle, Ohanian & Borsos, 1976; Muller-Eberhard, 1988; see Morgan, 1989, for references). There are two reasons for these effects. Firstly the plasma membrane contains proteins which inhibit the ability of C9 to lyse the cell, probably by blocking its insertion into the membrane (Davies *et al.*, 1989; 1990). Secondly, nucleated cells have mechanisms for protecting themselves against attack by removing the potentially lethal membrane attack complex via vesculation (Campbell & Luzio, 1981; Ramm *et al.*, 1983; Campbell & Morgan, 1985) (Table 4). It is here that we see particularly the value of using the Rubicon Hypothesis to relate the molecular biology to the cell biology, and where often there is a close interaction between intracellular Ca^{2+} and oxygen metabolites in the cell injury. The phenomena is well illustrated in neutrophils (Patel & Campbell, 1987; Fig. 3*a*,*b*) and oligodendrocytes attacked by complement (Scolding *et al.*, 1989*a*,*b*).

Measurement of cytosolic free Ca^{2+} using obelin, and fura 2 by single cell imaging, $O_2^{\overline{\cdot}}$ by chemiluminescence, and membrane permeability in single cells using the nuclear stain propidium iodide detected microscopi-

Table 4. *Cell responses associated with reversible damage membrane attack complex of complement*

Response	Cell example
1. Rise in cytosolic free Ca^{2+} through plasma membrane	neutrophil, macrophage platelet, oligodendrocyte
Release from internal store	neutrophil, oligodendrocyte
2. Reactive oxygen metabolite production and release	neutrophil macrophage renal cells synoviocytes
3. Prostaglandin release	synoviocytes oligodendrocytes neutrophils renal cells
4. Vesiculation (endocytosis and/or budding)	neutrophils oliogodendrocytes

Source:
For references see Campbell & Luzio (1981); Campbell (1983); Campbell & Morgan (1985); Scolding *et al.* (1989*a*, *b*); Morgan (1989).

cally or in the FACS, has shown that a nucleated cell can cross at least six thresholds or 'rubicons' following complement attack (Fig. 4). In erythrocytes and neutrophils rubicon I, a rise in cytosolic free Ca^{2+} through an increase in the permeability of the plasma membrane, can occur within 5 s of C9 binding to the membrane. However, in oligodendrocytes, where activation of complement occurs unusually by an antibody independent route involving a C1 binding protein on the cell surface, the Ca^{2+} rubicon may take up to 20 min to be crossed, after exposure to serum containing the complement. The membrane attack complex also appears to be able to activate the $P1P_2$ase (phospholipase C) which generates IP_3, since a release of Ca^{2+} from the internal store can also be detected (rubicon I). The rise in cytosolic Ca^{2+} activates Ca^{2+}-dependent processes (rubicon II), such as O_2^{-} production, within the cell (Hallett, Luzio & Campbell, 1981; Campbell & Morgan, 1985; Daniels *et al.*, 1990). Further evidence that the membrane attack complex is not a passive 'pore' in the membrane comes from studies on prostaglandin production. In neutrophils,

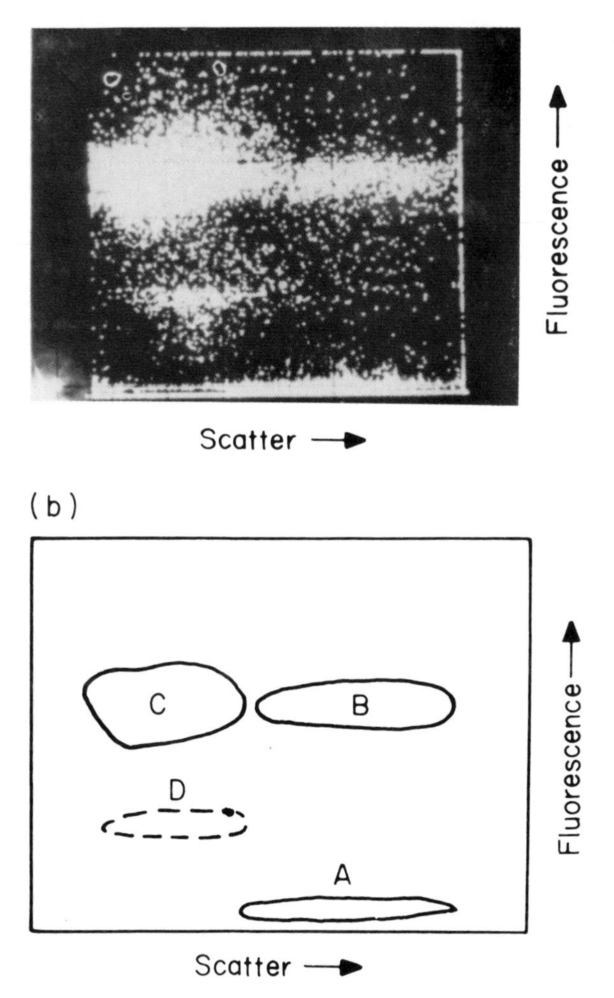

Fig. 3. Four sub-populations of neutrophils attacked by the MAC. Human neutrophils were incubated with the nuclear stain propidium iodide and attacked by antibody + 1/30 serum to form membrane attack complexes (MAC) on all of the cells. Four populations were identified and quantified using the fluorescence activated cell sorter. A = normal healthy cells; B = cells permeabilised to propidium iodide but capable of recovering from MAC attack; C = lysed cells; D = fragmented nuclei from lysed cells. The sequence for a cell undergoing six rubicons was thus A → B → C → D. From Patel & Campbell (1987); copyright *Immunology*, with permission.

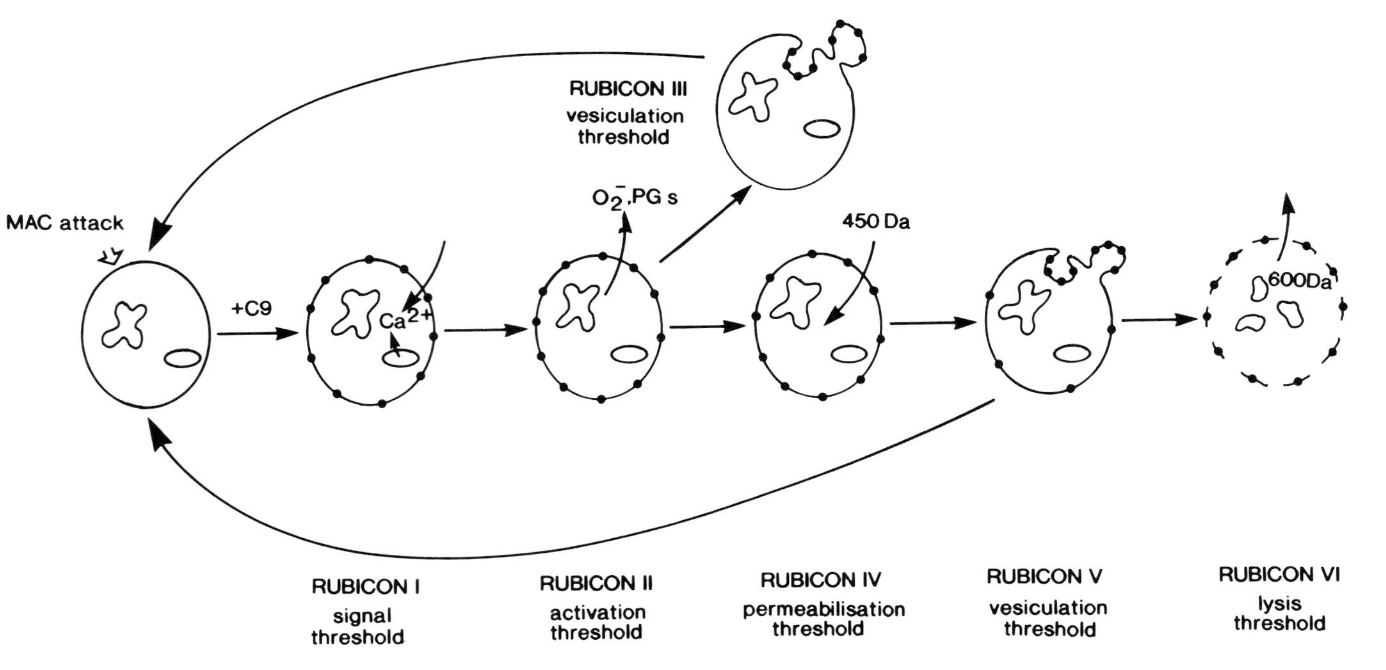

Fig. 4. Rubicon in defence against complement attack.

synoviocytes, kidney and several other cells, the membrane attack complexes (MAC) activate the eicosanoid-producing pathway (see Morgan, 1989, for references). The rise in intracellular Ca^{2+} also activates a protection mechanism, whereby the membrane attack complex is channelled into a vesicle which is removed from the cell by budding or endocytosis (rubicon III) (Campbell & Morgan, 1985; Morgan & Campbell, 1985; Scolding *et al.*, 1989*a*,*b*). Once the complexes have been removed the cell recovers rapidly to normal. Its ATP and free Ca^{2+} levels are similar to those of the original cells and it can again respond to cell stimuli. Some cells, however, cannot protect in time and cross the next rubicon (rubicon IV), an increase in the plasma membrane to molecules up to 450 Da (i.e. propidium iodide). There seems to be a critical size here, since fura 2 (550 Da) does not leak out at this stage, but does when the final threshold (rubicon VI) is crossed. Here the cell lyses and releases its contents. Between rubicons IV and VI the cells still have a chance to defend themselves (rubicon V). They can be isolated from the cell sorter (Fig. 3*b*). Following tissue culture they recover their ATP level and can again respond to stimuli. Those cells which cannot protect in time cross rubicon VI and die. The importance of using the Rubicon Hypothesis as a conceptual framework is seen when we try to relate the chemical events in the cells to the molecular biology of C9 (Stanley *et al.*, 1986), and in being able to predict the results of prolonging or shortening the period between any two rubicons.

C9 is a soluble, glycosylated protein, molecular weight 69 kDa, with a high *cys*-rich region involved in the polymerisation between C9 molecules. Our present working hypothesis for the molecular changes in C9 during membrane attack involves four steps, and is an extension of our original two-step model (Stanley *et al.*, 1986). Binding of C9 to C5b–8 and insertion in the plasma membrane causes rubicon I to be crossed, i.e. the movements of Ca^{2+}, and activates $P1P_2$ase. The Ca^{2+}, and probably at least one other intracellular signal, cause the cell to cross rubicon II (activation of Ca^{2+}-dependent processes in the cells) and rubicon III (the protection mechanism). The C9 molecules then begin to aggregate on the cell surface causing the cell to cross rubicon IV (permeability to propidium iodide). The intracellular changes resulting from this increase in permeability lead to a sudden, massive disruption of membrane structure which immediately lets out of the cell all the soluble constituents of the cytoplasm (rubicon VI; Fig. 4). Contrary to what was originally thought, polymerisation of C9 inactivates it (Dankert & Esser, 1987; Patel *et al.*, 1987*b*). Poly C9 is found in large amounts in vesicles budded off from the cell. What is not yet clear is whether the vesiculation which occurs between rubicons I and II (Ca^{2+}) and rubicon IV (permeability to 450

Da) requires polymerisation of C9. If it does then this simple model for the molecular biology of C9 is not quite correct since aggregation of C9 must occur prior to polymerisation.

A further consequence of the Rubicon Hypothesis is that it clarifies the understanding of the effects of inhibiting or enhancing the protection mechanism. This can occur in one of three ways.

1. Direct manipulation of the Ca^{2+} signal, e.g. using extracellular or intracellular EGTA.
2. Inducing the generation of another intracellular signal, e.g. cyclic AMP, which may or may not have a direct effect on the Ca^{2+} signal.
3. Inhibition of C9 polymerisation by α thrombin which cleaves C9 at his 285, but because of internal S–S bonds the two fragments C9a and C9b remain together and can still bind to C5b–8.

Inhibition of the rise in Ca^{2+} in neutrophils using EGTA (Morgan & Campbell, 1985) or a rise in cyclic AMP using adenosine (Roberts *et al.*, 1985 *a,b*), not only inhibits the protection mechanism but increases the *number* of cells which cross rubicon VI (lysis). α Thrombin-cleaved C9 is more potent at keeping the Ca^{2+} high in the cell and increases cell lysis. In contrast, in other cells (Boyle *et al.*, 1976) cyclic AMP may enhance the protection mechanism and thus reduce the number of lysed cells. These mechanisms have important implications for understanding the role of the MAC in inflammatory and immune-based diseases such as rheumatoid arthritis and multiple sclerosis.

Predictions

The ultimate test for any hypothesis is not simply whether it gives us a clearer understanding of observations already made, but rather whether it enables experiments to be designed from which really new observations are made. Such predictions arising from Rubicon fall into one of three groups.

1. The reclassification of phenomena in organs or cell populations, hitherto thought to occur via continuous, gradual changes in all the cells. When examined at the single cell level, these really involve individual cells crossing thresholds for signal production and end responses at different times after exposure to the stimulus, drug or pathogen.
2. Identification of the crucial molecular features of the

sequence from initiation at the plasma membrane to end response, or not, in an individual cell.

3. A molecular route for the appearance of phenomena during evolution. Only after a series of non-selectable steps (e.g. neutral mutations) is an event generated which is susceptible to the forces of Darwinian–Mendelian selection.

There are many phenomena in plant and animal cells where the biochemical processes appear continuous, for example, the production of $O_2^{\overline{\cdot}}$, the activation of intermediatry metabolism by hormones, the release of substrates through the plasma membrane, e.g. steroid hormones, the uptake and detoxification of drugs and toxins, the release of toxins. These need to be re-examined at the individual cell level to see whether a threshold exists, which, if not crossed, a particular cell will not exhibit the process seen in the whole population. It is also now necessary to examine membrane-pore-forming proteins such as T cell perforins, bacterial toxins, invertebrate toxins such as mellitin and others (Canicatti, 1990) and viral proteins to see if they induce the same series of rubicons as the membrane attack complex of complement, and in particular protection.

Any one of four mechanisms could be responsible for generating the threshold which causes the cell or organelle to cross a rubicon.

1. A molecular or atomic switch, e.g. the generation of an electronically excited state.
2. The formation and action of a molecular grouping e.g. a 'cap', a channel, a junction, or protein or enzyme complex.
3. The reaching of a critical concentration of a substrate or signal at the necessary location in the cell or organelle, via a 'cloud', wave or oscillation.
4. An organelle threshold, e.g. membrane–membrane fusion.

In *Chironomus* salivary gland a Ca^{2+} signal must reach the gap junction in order to switch off communication with a neighbouring cell (Fig. 5*a*; Rose & Loewenstein, 1976), and in the Medaka fish egg the Ca^{2+} wave initiated by one sperm is followed by a wave of $O_2^{\overline{\cdot}}$ production and enzyme secretion (Gilkey *et al.*, 1978). However, our recent results with neutrophils using fluorescent ratio imaging (Fig. 5B*c*) show that the time at which the rubicon for $O_2^{\overline{\cdot}}$ generation is crossed is dependent on the rubicon for Ca^{2+} signal generation, rather than on a difference in Ca^{2+} location within each cell (Hallett *et al.*, unpublished). This means that we now have to search for a molecular mechanism to explain the latency between receptor occupancy and the appearance of the Ca^{2+} signal.

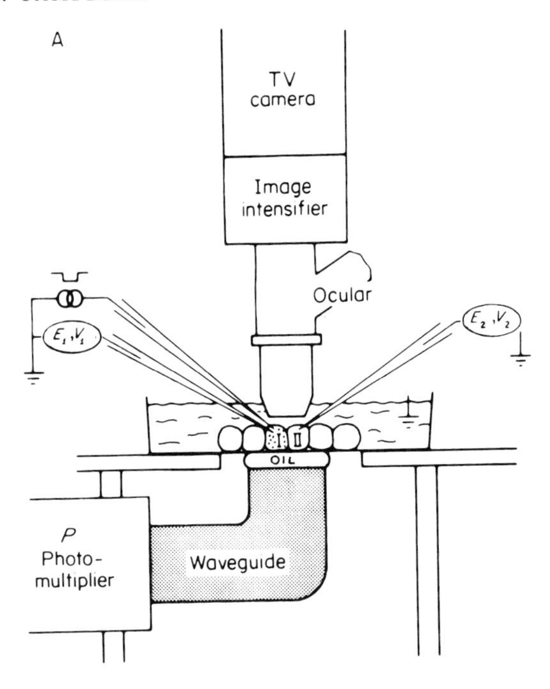

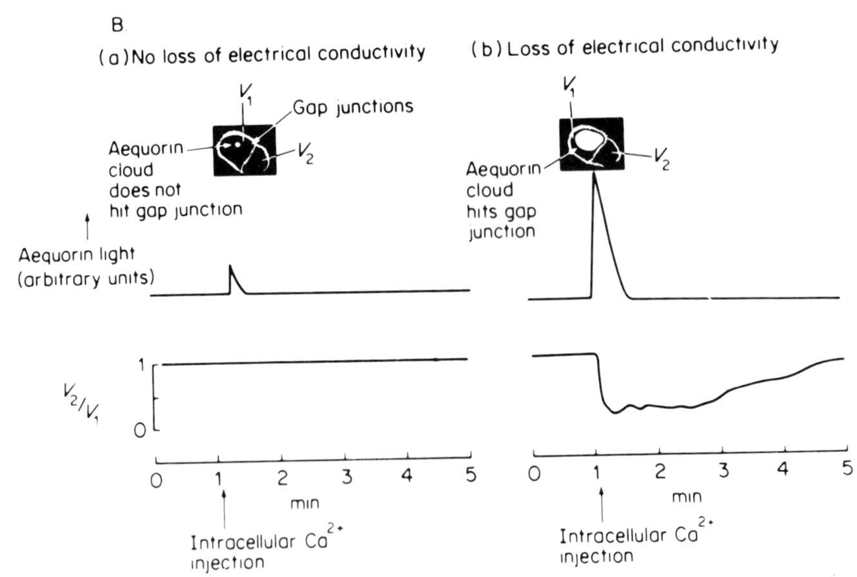

Fig. 5. Imaging of Ca^{2+} in single cells. (A) Apparatus. (B) Records: (*a*) and (*b*) blocking of gap junctions in *Chironomus* salivary gland from Campbell (1983); copyright Wiley and Sons, Chichester, with permission; (*c*) release of Ca^{2+} in a single neutrophil (Hallett, Davies & Campbell, unpublished).

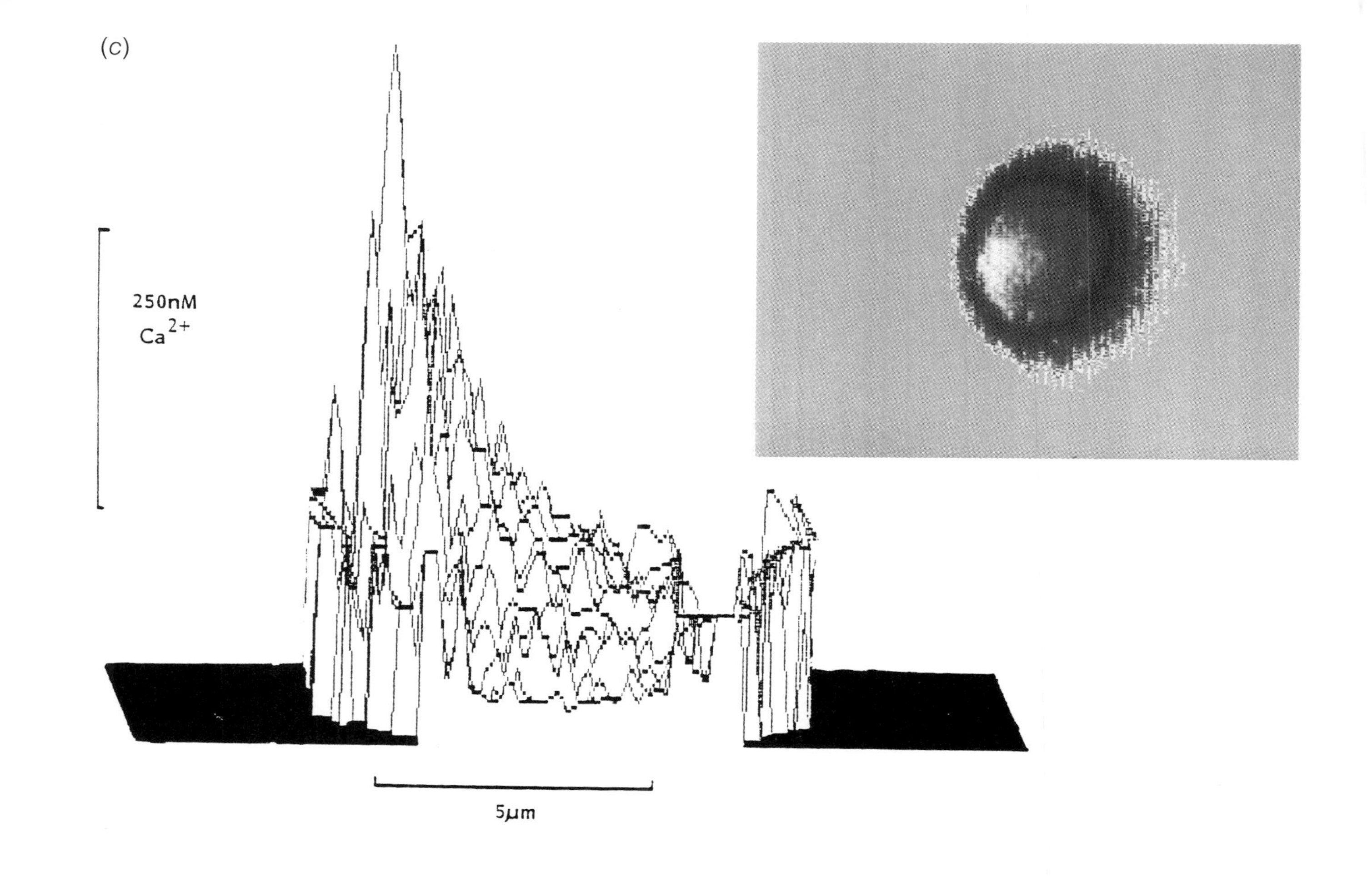
(c)
250nM
Ca2+
5µm

Three possibilities are the formation of a receptor cluster, the association of receptors with transducer molecules, e.g. G proteins, or the accumulation of the real primary signal either in the membrane or the cytosol which has to reach a critical concentration to open the Ca^{2+} 'channels' in the plasma membrane.

The need for new methodology

It is clear that in order to test the validity of the Rubicon Hypothesis and to follow its predictions, new methodology (see Table 5) and reagents are required to measure, locate and manipulate intracellular signals, energy phosphate (ATP and GTP), covalent modifications of proteins and the end response in single, living cells. The pioneering work of Ashley, Ridgway and others using aequorin to measure Ca^{2+} in 1970s (see Ashley & Campbell, 1979, for references) and of Tsien in the early 1980s developing ratio imaging (Poinie & Tsien, 1989; Williams & Fay, 1990) for fluors to measure and locate Ca^{2+}, and other ions has led to many important new findings, including the more recent discovery of oscillations in cytosolic free Ca^{2+} (Woods *et al.*, 1986; Berridge & Galione; Kanno *et al.*, 1989). But how are we to measure cyclic AMP, cyclic GMP, IP_3 and phosphorylation of proteins by protein kinase A, G, C or calmodulin in one cell?

Table 5. *Methods for chemical analysis of single, live cells*

1. Indicators
   - intracellular precipitates
   - absorbing dyes
   - fluors
   - bioluminescent proteins, mRNA and DNA
2. Incorporation into individual cells
   - micro-injection
   - fusion with vesicle or liposome
   - electroporation
   - mechanical permeabilisation (scrape loading)
   - transfection of cDNA
   - transgenic organism
3. Method of single cell analysis
   - whole cell response after isolation of each cell
   - cell sorting (FACS)
   - microscopical imaging

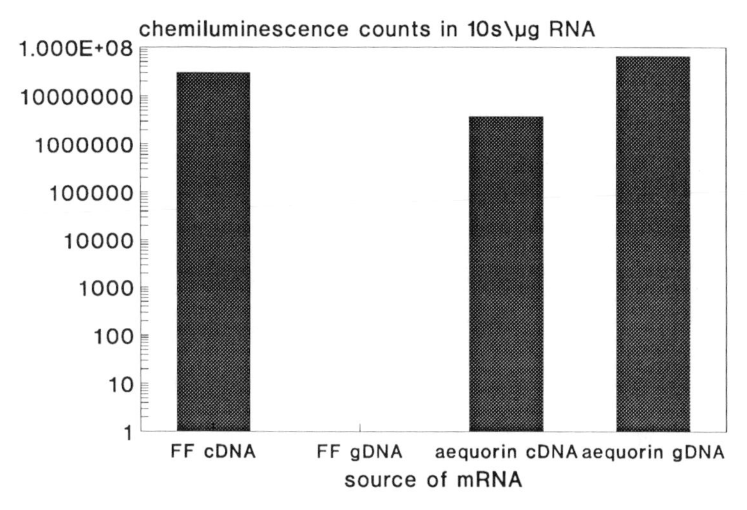

Fig. 6. Formation of bioluminescent proteins *in vitro* from cDNA using T7 RNA polymerase and rabbit reticulocyte lysate. cDNA coding to firefly luciferase or aequorin, or genomic DNA coding for aequorin, was amplified using the polymerase chain reaction (PCR), with a T7 promoter at the 5′ end. mRNA was formed and capped with m^7 GpppG *in vitro* using T7 RNA polymerase, and then translated using rabbit reticulocyte lysate to form active, light emitting protein. Data from Sala-Newby & Campbell, unpublished. For details of methods see Campbell *et al.* (1990) and Sala-Newby *et al.* (1990).

We have developed a strategy to achieve this by engineering bioluminescent proteins (Campbell, 1988; 1989*b*; Campbell *et al.*, 1990). The cDNAs coding for firefly luciferase and aequorin have been isolated (Sala-Newby, Kalsheker & Campbell, 1990) and a T7 RNA polymerase promoter added to the 5′ end using the polymerase chain reaction (PCR). This enables active light-emitting protein to be produced *in vitro* using T7 RNA polymerase, followed by translation of the mRNA formed in rabbit reticulocyte lysate (Fig. 6). A protein kinase A site has been engineered into these light-emitting proteins using PCR such that phosphorylation will introduce a change in intensity and/or colour in the light emission (Jenkins, Sala-Newby & Campbell, 1990). The controlled expression of these unique indicator cDNAs will enable, for first time, signals and protein phosphorylation to be measured not only in the cells but in whole organs.

The horizon

In some respects the Rubicon Hypothesis is a more precise way of describing the pleiotypic response of cells highlighted by Tomkins (Hershko *et al.*, 1971). However, it differs in one crucial respect, namely the principle of quantal leaps in the chemistry and physics of each cell, and their timing. Rubicon focuses attention on how essential it is to look at each cell one at a time if real progress is to be made in unravelling the molecular basis of cell activation and injury. It is obvious that the age and past history of cells in a population will mean that it is inevitable that there is heterogeneity in the timing and magnitude of individual cell responses which make up the population. What is perhaps less obvious, and is brought to the fore by Rubicon, is that the molecular pathway necessary to take the cell from initiation to end response involves a series of quantal leaps. What is needed now is to identify the combination of chemical reactions and physical changes which must precede each leap if the correct sequence is to be followed. A further, perhaps more elusive, goal is to discover a precise, simple mathematical relationship which might help to resolve a puzzle which has existed since Lamarck and others first used the word 'biology': is there something special about the chemistry and physics of a cell, which distinguishes it from 'non-living' processes, and allows us to call it 'alive'? Only when we know the answer to this question will we truly understand where physiology ends and pathology begins.

Acknowledgements

I thank the MRC, the AFRC, the Arthritis and Rheumatism Council, and the Multiple Sclerosis Society for financial support, and many of my colleagues including in particular the past and present members of my research group, for their experiments and hard work, Dr J. Paul Luzio and Dr Gerry V. Brenchley for their friendship and many stimulating discussions, and finally my wife Dr Stephanie Matthews for allowing me to enjoy the inspiration of our cottage in Ynys Mon, where I wrote this article.

References

Akam, M. (1987). The molecular basis for metameric pattern in the Drosophila embryo. *Development* **101**, 1–22.

Ashley, C.C., Campbell, A.K. (eds) (1979). *The Detection and Measurement of Free Ca^{2+} in Cells*, pp. 461. Amsterdam: Elsevier/North Holland.

Baerlocher, F. (1990). The Gaia hypothesis: a fruitful fallacy. *Experientia* **46**, 232–8.

Berridge, M.J. & Galione, A. (1988). Cytosolic calcium oscillators. *Federation of the American Society for Experimental Biology Journal* **2**, 3074–82.

Berridge, M.J. & Irvine, R. (1984). Inositol trisphosphate, a novel second messenger in cellular signal transduction. *Nature* **312**, 315–21.

Berridge, M.J. & Irvine, R. (1990). Inositol phosphate and cell signalling. *Nature* **341**, 197–205.

Boyle, M.D.P., Ohanian, S.H. & Borsos, T. (1976). Studies on the terminal stages of antibody–complement mediated killing of a tumour cell. II. Inhibition of transformation of T* to dead cells by 3′5′ cAMP. *Journal of Immunology* **116**, 1276–9.

Campbell, A.K. (1983). *Intracellular Calcium: Its Universal Role as Regulator*, pp. 556. Chichester: John Wiley and Sons.

Campbell, A.K. (1987). Intracellular calcium: friend or foe? *Clinical Science* **72**, 1–10.

Campbell, A.K. (1988). *Chemiluminescence: Principles and Applications in Biology and Medicine*, p. 608. Chichester and Weinheim: Horwood/VCH.

Campbell, A.K. (1989*a*). In *Methodological Surveys in Biochemistry and Analysis*, vol. 19, Biochemical Approaches to Cellular Calcium, ed. E. Reid, G.M.W. Cook & J.P. Luzio, pp. 1–14. London: Royal Society of Chemistry. (A century of intracellular calcium.)

Campbell, A.K. (1989*b*). *British Patent Application 8916806.6*. Bioluminescent proteins.

Campbell, A.K. & Hallett, M.B. (1983). Measurement of intracellular calcium ions and oxygen radicals in polymorphonuclear leucocyte-erythrocyte ghost hybrids. *Journal of Physiology* **338**, 537–62.

Campbell, A.K. & Luzio, J.P. (1981). Intracellular free calcium as a pathogen in cell damage initiated by the immune system. *Experientia* **37**, 1110–12.

Campbell, A.K. & Morgan, B.P. (1985). Monoclonal antibodies demonstrate protection of polymorphonuclear leucocytes against complement attack. *Nature* **317**, 164–6.

Campbell, A.K., Patel, A.K., Razavi, Z.S. & McCapra, F. (1988). Formation of the Ca^{2+}-activated photoprotein from obelin and mRNA inside human neutrophils. *Biochemical Journal* **252**, 143–9.

Campbell, A.K., Sala-Newby, G., Aston, P., Jenkins, T. & Kalshekar, N. (1990). From Luc and Phot genes to hospital bed. *Journal of Bioluminescence and Chemiluminescence* **4**, 131–9.

Canicatti, C. (1990). Hemolysins: pore-forming proteins in invertebrates. *Experientia* **46**, 239–44.

Carafoli, E. (1987). Intracellular calcium homeostasis. *Annual Review of Biochemistry* **56**, 395–433.

Cohen, P. (1988). Protein phosphorylation and hormone action. *Proceedings of the Royal Society of London* B**234**, 115–44.

Daniels, R.H., Houston, W.A.J., Petersen, M.M., Williams, J.D., Williams, B.D. & Morgan, B.P. (1990). Stimulation of human rheumatoid synovial cells by non-lethal complement membrane attack. *Immunology* **69**, 237–42.

Dankert, J.R. & Esser, A.F. (1987). Bacterial killing by complement: C-9 killing in the absence of C5b–8. *Biochemical Journal* **244**, 393–9.

Dankert, J.R., Shiver, J.W. & Esser, A.F. (1985). Ninth component of complement: self aggregation and interaction with lipids. *Biochemistry* **24**, 2754–60.

Davies, A., Hallett, M.B. & Campbell, A.K. (1991). Single cell imaging demonstrates release of calcium from a single store in human neutrophils. *Immunology* (in press).

Davies, A., Simmons, D.L., Hale, G., Harrison, R.A., Tighe, H., Lachmann, P.J. & Waldmann, H. (1989). CD59, an LY-6 like protein expressed in human lymphoid cells, regulates the action of the complement membrane attack complex on homologous cells. *Journal of Experimental Medicine* **170**, 637–54.

Davies, A., Watts, M.J. & Morgan, B.P. (1990). Control of complement membrane attack. In *Complement*, ed. R.A. Harrison. Cambridge University Press (in press).

Duncan, C.J. (ed.) (1976). *Calcium in Biological Systems*, Symposium Society of Experimental Biology **30**, 1–500.

Edwards, S.W., Morgan, B.P., Hoy, T.G., Luzio, J.P. & Campbell, A.K. (1983). Complement mediated lysis of pigeon erythrocyte 'ghosts' analysed by flow cytometry: evidence for the involvement of a threshold phenomenon. *Biochemical Journal* **216**, 195–202.

Gilkey, J.C., Jaffe, L., Ridgway, E.B. & Reynolds, G.T. (1978). A free calcium wave traverses the activating egg of the medaka, Oryzias latipes. *Journal of Cell Biology* **76**, 448–66.

Gilman, A.G. (1987). G Proteins: transducer of receptor-generated signals. *Annual Review of Biochemistry* **56**, 615–49.

Goldberg, B. & Green, H. (1959). Effect of antibody and complement on permeability control in ascites tumour cells and erthrocytes. *Journal of Experimental Medicine* **109**, 505–10.

Green, G., Barrow, P. & Goldberg, B. (1959). The cytotoxic action of immune gamma globulins and complement on Krebs ascites tumour cells. *Journal of Experimental Medicine* **110**, 689–713.

Hallett, M.B. & Campbell, A.K. (1982). Measurement of changes in cytoplasmic free Ca in fused cell hybrids. *Nature* **295**, 155–8.

Hallett, M.B., Davies, A. & Campbell, A.K. (1990). Oxidase activation in individual neutrophils is dependent on the onset and magnitude of the Ca^{2+} signal. *Cell Calcium* **11**, 655–63.

Hallett, M.B., Luzio, J.P. & Campbell, A.K. (1981). Stimulation of Ca-dependent chemiluminescence in rat polymorphonuclear leucocytes by polystyrene beads and non-lytic action of complement. *Immunology* **434**, 569–76.

Halliwell, B. & Gutteridge, J.M.C. (1989). *Free Radicals in Biology and Medicine*, 2nd edn, pp. 1–346. Oxford: Clarendon Press.

Heilbrunn, L.V. (1937). *An Outline of General Physiology*, 1st edn. Philadelphia: Saunders.

Heilbrunn, L.V. (1956). The Dynamics of Living Protoplasm. New York: Academic Press.

Hershko, A.P., Mamon, R., Shields, R. & Tomkins, G.M. (1971). Pleitypic response. *Nature* **232**, 206–11.

Jenkins, T., Sala-Newby, G. & Campbell, A.K. (1990). Measurement of protein phosphorylation by covalent modification of firefly luciferase. *Biochemistry Society Transactions* **18**, 463–5.

Kanno, T. Saito, T. & Yamashita, T. (1989). Spatial and temporal oscillation of [Ca^{2+}] during continuous stimulation with CCK-8 in isolated rat pancreatic acini. *Biomedical Research* **10**, 475–84.

Krustal, B.A. & Maxfield, F.R. (1987). Cytosolic free calcium increases before and oscillates during frustrated phagocytosis in macrophages. *Journal of Cell Biology* **105**, 2685–93.

Krustal, B.A., Shak, S. & Maxfield, F.R. (1986). Spreading of human neutrophils is immediately preceded by a large increase in cytoplasmic free calcium. *Proceedings of the National Academy of Science USA* **83**, 2919–23.

Lew, D.P., Monod, A., Waldvogel, F.A. & Pozzan, T. (1987). Role of cytosolic free calcium and phospholipase C in leukotriene-B4-stimulated secretion in human neutrophils. *European Journal of Biochemistry* **162**, 161–8.

Lovelock, J. (1988). *The Ages of Gaia: Biography of our Living Earth*, pp. 1–252. Oxford University Press.

Marks, P.W. & Maxfield, F.R. (1990). Local and global changes in cytosolic free calcium in neutrophils during chemotaxis and phargocytosis. *Cell Calcium* **11**, 181–90.

Mitchell, P. (1966). Chemiosmotic coupling in oxidative and photosynthetic phosphorylation. *Biological Reviews* **41**, 445–502.

Morgan, B.P. (1989). Complement membrane attack on nucleated cells: resistance, recovery and non-lethal effects. *Biochemical Journal* **264**, 1–14.

Morgan, B.P. & Campbell, A.K. (1985). The recovery of human polymorphonuclear leucocytes from sublytic complement attack is mediated by changes in intracellular free calcium. *Biochemical Journal* **231**, 205–8.

Muller-Eberhard, H.J. (1988). Molecular organization and function of the complement system. *Annual Review of Biochemistry* **57**, 321–47.

Patel, A. & Campbell, A.K. (1987). The membrane attack complex of complement induces permeability changes via thresholds in individual cells. *Immunology* **60**, 135–46.

Patel, A., Hallett, M.B. & Campbell, A.K. (1987*a*). Threshold responses in reactive oxygen metabolite production individual neutrophils, detected by flow cytometry and microfluorimetry. *Biochemical Journal* **248**, 173–80.

Patel, A., Morgan, B.P. & Campbell, A.K. (1987*b*). The ring-like classical complement lesion is not the functional pore of the membrane attack complex. *Biochemical Society Transactions* **15**, 659–60.

Persechini, A., Moncrief, N.D. & Kretsinger, R.H. (1989). The EF-hand family of calcium-modulated proteins. *Trends in Neurosciences* **12**, 462–7.

Poinie, M. & Tsien, R.Y. (1989). Fluorescence ratio imaging: a new window into intracellular signalling. *Trends in Biochemical Sciences* **11**, 450–5.

Ramm, L.E., Whitlow, M.B., Koski, C.L., Shin, M.L. & Mayer, M.M. (1983). Elimination of complement channels from the plasma membrane of U937, a nucleated mammalian cell line. *Journal of Immunology* **131**, 1411–16.

Rasmussen, H. (1989). The cycling of calcium as an intracellular messenger. *Scientific American*, pp. 44–51.

Reid, E., Cook, G.M.W. & Luzio, J.P. (eds) (1989). *Methodological Surveys in Biochemical Approaches to Cellular Calcium*, pp. 1–495. London: Royal Society of Chemistry.

Roberts P.A., Knight, J. & Campbell, A.K. (1987). Pholasin – a bioluminescent indicator for detecting activation of single neutrophils. *Analytical Biochemistry* **160**, 139–48.

Roberts, P.A., Morgan, B.P. & Campbell, A.K. (1985*b*). 2-Cadenosine inhibits complement-induced reactive oxygen metabolite production and recovery of human polymorphonuclear leucocytes attacked by complement. *Biochemical Biophysical Research Communications* **126**, 692–7.

Roberts, P.A., Newby, A.C., Hallett, M.B. & Campbell, A.K. (1985*a*). Inhibition by adenosine of reactive oxygen metabolite production by human polymorphonuclear leucocytes. *Biochemical Journal* **227**, 669–74.

Rose, B. & Loewenstein, W.R. (1976). Permeability of a cell junction and the local cytoplasmic free ionised calcium concentration: a study with aequorin. *Journal of Membrane Biology* **28**, 87–119.

Sala-Newby, G., Kalsheker, N. & Campbell, A.K. (1990). In vitro production of translatable firefly luciferase mRNA from cloned cDNA. *Biochemical Society Transactions* **18**, 459–60.

Scolding, N.J., Houston, W.A.J., Morgan, B.P., Campbell, A.K. & Compston, D.A.S. (1989*a*). Reversible injury of cultured rat oligodendrocytes by complement. *Immunology* **67**, 441–6.

Scolding, N.J., Morgan, B.P., Houston, W.A.J., Linington, C., Campbell, A.K. & Compston, D.A.S. (1989*b*). Oligodendrocytes activate complement but resist lysis by vesicular removal of membrane attack complexes. *Nature* **339**, 690–2.

Stanley, K.K., Page, M., Campbell, A.K. & Luzio, J.P. (1986). A mechanism for the insertion of complement component C9 into target membranes. *Molecular Immunology* **23**, 451–8.

Volpe, P., Krause, K.H., Hashimotot, S., Zorzato, F., Pozzan, T., Meldolesi, J. & Lew, D.P. (1988). 'Calciosome', a cytoplasmic organelle: the inositol 1,4,5-trisphosphate-sensitive Ca^2 store of nonmuscle cells? *Proceedings of the National Academy of Science USA* **85**, 1091–5.

Williams, D.A. & Fay, F.S. (eds) (1990). Imaging of cell calcium. *Cell Calcium* **11**, 55–249.

Woods, N.M., Cuthbertson, K.S.R. & Cobbold, P.H. (1986). Repetitive transient rises in cytoplasmic free Ca in hormone-stimulated hepatocytes. *Nature* **319**, 600–2.

Index

Page numbers in *italic* type refer to tables or figures

A23187 1, 2, 19, 24, 102, 106, 140, 159, 174, 175
acetylcholine 125, 196
  receptors 3
acetylcholinesterase 116, *117*
acid hydrolases
  inhibitors 159
actin filaments
  dissolution 149
action potential *197*
adenine nucleotides 165
adenosine triphosphate (ATP) 3, 5, 24, 97, 167, 168, *181*, 181, 210
  ischaemia 173
  synthesis 127
adenylate cyclase 116, *117*, 179, 193
adenylate kinase 116, *117*
adriamycin 26, 172
aequorin 210, 211
ageing 17
  muscle 159
alkoxy radicals 168
allopurinol 106, 182
  inhibition of xanthine oxidase 174
amiloride *103*, 105, *107*
  inhibition of Na^+/H^+ antiporter 102
anoxia 98, *99*, 101, *103*, 104, 106, *107*, *155*, 159
anthracycline quinones
  cardiac toxicity 109
  one-electron reduction 109
apoptosis 23, *197*
  thymocytes 24
arachidonic acid 3, 4, 12, 49, 78, 79, 141, 144, 175, 182
  cycloxygenase 103
  lipoxygenase 103
  neutrophils 48
  phospholipase A_2 103
arrhythmias 88, *89*
arrhythmogenic 87
arthritis 193
ascorbic acid
  see also vitamin C 78, 166
atheroma 193
atomic absorption spectroscopy 169–70

BAPTA 25
B cell 191
benzoylperoxide 25
Blebs 18, 19, 20

C1 binding protein 202
C9 201, 202, 206
  molecular weight 205
  polymerisation 206
caesium
  potassium replaced with 84
caffeine 82, *103*, 104, 105, 109, 128
  release of Ca^{2+} from SR 102
  sarcoplasmic reticulum 81
calcium
  cell damage 97
  damage to normal skeletal muscle 140
  determinants of free-radical-mediated processes 167
  electrochemical gradient 125
  free-radical-mediated post-ischaemic injury 174
  kidney damage 173
  oscillations in cytosolic free Ca^{2+} 210
  phospholipase A_2 98, 160
calcium-activated neutral proteases 4
calcium antagonists 140
  skeletal muscle 141
calcium-ATPases
  endoplasmic reticulum 124
  increased sensitivity to halothane 128
  plasma membrane 124
  sarcolemma 125, *126*, 127

sarcoplasmic reticulum 125,127
calcium-dependent cytotoxin mechanisms 22–4
calcium-dependent endonuclease 24
calcium-dependent proteases 23, 141
calcium homeostasis 2, 17, 20, 109, 116, 129, 174, 181
  and cardiac muscle 82
calcium influx 141
calcium, intracellular 140
  mdx mouse 144
  rise 101
  triggers contraction 97
calcium inward current 83–5
  Rose bengal 82
calcium metabolism
  malignant hyperthermia 120
  porcine stress syndrome 120
calcium overload 81, 85, 89, *90*
calcium paradox 3, 98, *99*, 101, 102, 103, 104, 105
  oxygen radicals 100–1
  stages 100
calcium pumps, 173, 181
calcium-sensitive microelectrodes 124
calcium transport 101
calcium uptake 97
calmodulin 39, 125, 193, 210
calpain 22, 23
  cytoskeleton-associated proteins 20
  proteases 104
calsequestrin 127
cAMP 161, *192*, 193, 206, 210
5-carbadeaza-FAD
  flavin analogues 41
carbon tetrachloride 2, 3, 11
  hepatotoxicity 14
cardiac arrhythmia 80
cardiac muscle
  leukotrienes 103
  prostaglandins 103
cardiac myocytes 151, *153*, *154*
catalase 2, *19*, 78, 80, 98, 101, 106, 166
catecholamines
  auto-oxidation 78
cathepsin G 36
cGMP *192*, 210
chemotaxis 191, *197*
Chironomus
  salivary gland 207, *208*
cholinesterases 3
chronic granulomatous disease (CGD) 35, 39, 41, 49, 50
CI 106, *107*
complement 201
  attack *204*
  oligodendrocytes 201
  reversible damage *202*
conjugated dienes 120
contractile activity 140
copper 12, 13, 37
creatine kinase (CK) *99*, *103*, 106, *107*, 109, 120, 122, 123, 143
  isoform pattern 139
  release 100, 101, 104, 110
cyanide 3, 35
cyclic nucleotides 191
cycloheximide 53, *58*
cycloxgenase 3, 12, 48, 98, 168, 175
cystic fobrosis 193
cytochalasin D 19
cytochrome *b* 40, 41, 51
  absorption maximum 40
  NADPH oxidase 40
cytochrome oxidase 166
cytokines 58, 59
cytoskeletal proteins
  calcium-activated proteases 23
cytoskeleton *198*
cytosolic calcium *130*

dantrolene sodium 2
defensins 36
desferrioxamine 80, 101, 106, 167, 168, 169, *170*, *172*
detergents 108, 140
diacylglycerol (DAG) 39, 46, 47, 48, 51, 176, 191, *192*, 193
dibucaine 175
  inhibition of phospholipase A_2 175
2,7-dichlorofluorescin 197, *199*
dihydropyridine 125, *126*, 127
diisopropylfluorophosphate 3
diltiazem 3
dimethylthiourea 101, 106
2,3-dimethoxy-1,4-naphthquinone 26
5,5-dimethyl-1-pyrroline-1-oxide (DMPO) 37
2,4-dinitrophenol 6, 102, 141
dipalmitoyl-phosphatidylcholine 150
dodecyl imidazole 151, *154*, *155*, *156*, *157*, *158*
  swelling of SR 159
DPI 41
Duchenne muscular dystrophy 1, 4
  biceps muscle 144
  dystrophin 143
  elevated release of creatine kinase 144
  prostaglandin E_2 144
dystrophic muscle 159

dystrophic muscle (*cont.*)
calcium 139
free radicals 139
dystrophin 1, 143
Becker muscular dystrophy 143
Duchenne muscular dystrophy 143

Eicosanoid cascade 3, 4
eicosanoid production 180, 182, 205
elastase 36
electron paramagnetic resonance spectroscopy 124
electron spin resonance 80
from skeletal muscle 142
electrons 108, 109
transferance to membrane proteins 101
endocytosis *202*, 205
endoplasmic reticulum 2, 3, 165, *198*
Ca^{2+}-sequestering 21
ethane 120
excitation–contraction coupling 82, 124
exercise 139, 142
plasma CK 142

ferritin 172
fertilisation *197*
firefly luciferase 211
flavoprotein inhibitor, DPI 41
fluorescent ratio imaging 207
fMet–Leu–Phe 45, *46*, *47*, 48, 51, 52, 54, 56, 59
free fatty acids
acting as detergents 141
free radicals 11–15, 118, *130*
damage to muscle cells 142
malignant hyperthermia 119, 120
porcine stress syndrome 119, 120
vitamin E deficiency 119
see also oxygen radicals
freeze fracture 99
fura 2 200, 201, 205

gap junction *198*, 207
gas liquid chromatography 175
gluconeogenesis 178
glucose-6-phosphate dehydrogenase 116, *117*
glucose phosphate isomerase
locus 129
glutathione 18, 80, 121, 122, 166
as a electron donor 18
GSSG 27
intracellular pool 18
glutathione peroxidase 2, *19*, 78, 116, *117*, 119, 121
glutathione reductase 18, *19*
Golgi zone 160
G-protein 39, 45, 193, 210
GTP 210
guanylate cyclase 179, 193

H^+ channel 108
H^+ efflux 100, 102, *103*, 106, 108
Haber–Weiss reaction 12, 37, 167
haemoglobin 201
breakdown 78
halothane 2, 115, 118, 120, *121*, 128, 129, *130*
lipid peroxidation 121
metabolism in endoplasmic reticulum 118
neutrophils 109
protein kinase C 109
superoxide production 109
hamster *157*, 158
skeletal muscle 140
heart 2, 3
damage detected 98
guinea pig 101
rabbit 101
hepatocytes 2, 4, 5
high-energy phosphates 104
HPLC 179
hydrogen peroxide 17, 18, 25, 37, 39, 60, 78, 106, 166, 197
hydroperoxide peroxidase 119
hydroxyl radical 12–13, 17, 78, 106, 118, 166, 167
4-hydroxynonenal 179
17-hydroxywortmannin 48
hyperbaric oxygen 24
hypercontraction 97
hyperosmolar solutions 108
hypoxanthine 167, *181*
hypoxia 172, *178*, 179
damage to cardiac tissue 140

indomethacin 175, 182
inositol bisphosphate (IP_2) 179
inositol monophosphate (IP_1) 179
inositol 1, 3, 4, 5-tetrakisphosphate *39*, *46*
inositol 1, 4, 5-triphosphate (IP_3) 39, 45, 127, 176, 179, 193, 202, 210
insulin secretion 191
γ-interferon 52, 58, 59
ion channels 81
ion exchangers 81
iron 12, 13, 37, 168, 169, *170*, *171*, *172*,

172, 173, 176, 182
determinants of free-radical-mediated processes 167
role in kidney 167–73
ischaemic myocardium 77–91

kidney 2,
storage of 165

lamellar bodies 161
Langendorff-perfused heart 98
lattice myelin 161
leucine methylester 150, 151, *152*, *153*, *154*, *155*
swelling of SR 159
leukotriene B_4 144
leukotrienes 3, 4, 48, 97, 98, 175, *181*, 182
leupeptin 4
lipid bodies *152*, *153*, *154*, *155*, *156*, *157*, *158*, 159, 161
danaged muscle cells 151
mitochondria 149, 150, 151
myelin-like figures 151
store of phospholipids 160
lipid peroxidation 13, 14, 118, 120, 142, 166, 167, 168, 171, *172*, 172, 174, 176, *178*, 178, 179, *181*
initiated by Fe^{2+} 124
lipoxygenase 3, 12, 48, 98, 175
inhibitors 4
lysis *198*
lysolecithin 3, 4, 5
lysophosphatides 175, 176, *177*
intracellular accumulation 80
lysophospholipids 141
lysosomal acid hydrolases 104
lysosomal activity
endocytosis 159
lysosomal apparatus 159
lysosomal cathepsins 150, 159
lysosomal hydrolases 97
lysosomal proteases 141
lysosomes 4
lysosomal enzymes 5, 97, 104, 141, 150, 159, 165
and prostaglandins 5
sarcotubular system 150
skeletal muscle 150

Magnesium
activation by 49
calcium-ATPase pump 125
malignant hyperthermia 1, 2, 4, 14, 115–37
calcium metabolism 120
calcium regulation 128, 129
defect in cell antioxidant systems 129
free radicals 119, 120
Islets of Langerhans 124
lymphocytes 124
membrane abnormalities 122, 129
red cell membrane 124
malondialdehyde 142, *178*
mannitol 80, 101, 167
mdx mouse 1
creatine kinase 143
intracellular calcium 144
lack of dystrophin 143
model of muscular dystrophy 143
Medaka egg 207
mellitin 207
membrane perturbation 108
membrane phospholipids 99, 141, 165
membrane potential
oscillations *90*
membrane pumps
damage 81
ion channels 81
ion exchangers 81
menadione 18, 21, 26
blebbing 19
cellular damage 109
creatine kinase release 109
platelets 23
protein thiols 18
mepacrine
calcium paradox 104
metabolic acidosis 115
microcystin-LR 19
mitochondria *155*
bars 151, *152*, *153*, 158
calcium overload *130*
calcium pool 127
calcium sequestering 21, 97, 141, 160
cardiac muscle cells 149
cristae 151, *152*, *158*
damaged muscle *152*
dinitrophenol 102
dysfunction 5
hydrogen peroxide 78
hydroxyl radical 78
impaired function in kidney 165
increase in number 151
inhibitors 141
injury 167
injury by calcium overload 175
internal septa 97, 151, *152*, 158
lipid bodies *153*
manganese-containing form 119

mitochondria (*cont.*)
myelin figures *154*
oxidative energy production 141
phospholipase A_2 5, 161
septation 97
skeletal muscle 149
subdivided *156*
superoxide anion 78
superoxide dismutase 13, 160
superoxide production 12, 17, 160
swelling 151, *153*, *156*, 159
ultrastructure 149, 158
uniporter *126*
mitochondrial division *198*
intracellular calcium 150
mouse
cardiac cells 102
multiple sclerosis 4, 193
multivesicular bodies 160
muscular dystrophy 193
see also Duchenne muscular dystrophy 193
myelin-like figures 151, *154*, 159, 160
in mitochondria 150, *153*, 158
myeloperoxidase 36, 37, 39
myocytes 80, 81, *83*, 84, *87*
myofilament apparatus 97
myoglobin breakdown 78
myophosphorylase B 116, *117*

Na^+/H^+ antiporter 100, 105, 106, 107, 108, 109, 125, *126*
amiloride 102
mitochondrial 127
Na/K-ATPase activity
inhibition of 86
in isolated hearts 85
membrane vesicles 85
NADPH 18
NADPH oxidase 11, 39–40, 60, 78
activation 42, 49
adhesion receptors 44
assembly 49
components 40, 42
cytochrome b 40
flavoproteins 41
IgG Fc receptors 43
receptors 42
respiratory burst 36
and superoxide 11, 12
NAD(P)H dehydrogenase 106
NAD(P)H oxido-reductase 108, 110
electrons 109
generates H^+ 109
Na^+/H^+ antiporter 109
nematocyst *197*
neuromuscular junction 3
neutrophil 1, 35–60, 108, *191*, *197*, *202*
arachidonic acid 48
cytokines 58
gene expression 53, 58, 59
phospholipase A_2 48
protein kinase C 46
role of Ca^{2+} 46
N formyl met-leu-phe (Nfmlp) 198, *199* 199
Nigleria 97
nitrogen 108, 109, 110
nordihydroguaretic acid
calcium paradox 104
lipoxygenase inhibitor 103
nuclear magnetic resonance 124

obelin 199, 201
oligodendrocyte *197*, 201, *202*, 202
oncogenes 193
orinthine decarboxylase 26
oscillatory calcium release 82
ouabain 88
oxidant stress 81, *83*, 88
calcium channel 85
calcium overload 85, 86
cardiac arrhythmias 79
conductance *89*
free radicals 80
inhibition of Na/K-ATPase 86
ionic currents 82, *87*
membrane electrophysiology 91
membrane potential 81
Na-pump current 89
photoactivation of rose bengal *84*, 87
SR function 82
oxidative phosphorylation 127, *130*
oxygen 108, 109
oxygen metabolites 110
oxygen paradox 3, 98, *99*, 104, 105, 109
oxygen radicals 98, 108, 110, *181*
generation 109
ozone 25

pale soft exudative (PSE) 115
Paramecium 191
parathyroidectomy 140
pentane
excretion 142
in vitamin E deficiency 120
peroxidases 166
peroxy radicals 12, 168
pH_i 4, 97, 102, 165, 172
and acid hydrolases 5, 6

phalloidin 19
α-phenyl *N-tert*-butyl nitrone 80
phorgol ester 198, 199
phorbol myristate acetate (PMA) 46, 48, 52, 54, 56, 161, 198
phosphatidic acid 48
phosphatidylcholine 160
phosphatidylinositides 173
phosphatidylinositols (PIP_2) 176, 179, *180*, 180
phosphatidylinositol-4,5-biphosphate 45
phospholipase 144, 175, 176, 182, 193
phospholipase A_2 3, 4, 5, 39, 50, 127, *130*, 141, 160, 175, 176, *177*, 179
  arachidonic acid 103
  calcium 23, 98
  calcium activation 103
  calmodulin 23
  mepacrine 103
  mitochondrial 160, 161
  neutrophils 48
phospholipase C 39, 50, 173, 180, 202
phospholipase D 39, 50
  wartmannin 48
phototaxis *197*
plannar lipid bilayers 82
poly C9 205
porcine stress syndrome (PSS) 115, 119, 120
  calcium regulation 128, 129
  defect in cell antioxidant systems 129
  enzyme/protein abnormalities 116
  GSH 121
  membrane abnormalities 122, 129
  similarities with vitamin E deficiency 122
porcine stress syndrome-resistant pig
  creatine kinase *123*
  Islets of Langerhans 124
  lymphocytes 124
  membrane defect 123
  pyruvate kinase *123*
  red cell membrane 124
positive feedback 103
potassium
  gradients 80
  washout 80
propidium iodide 201, *203*, 205
prostacyclin 175
  analogues 182
prostaglandins 3, 4, 48, 97, 98, 141, *181*, *202*, 202
  and intralysosomal proteolysis 5
protein kinase A 161, 210, 211
protein kinase C 2, 4, 25, 26, 27, 39, 46, 47, 50, 51, 100, *103*, 105, 106, *107*, 108, 109, 161, 176, 210
protein kinase G 210
protein thiols 21
pulmonary surfactant 160–1
  lipid bodies 160
  myelin figures 160
pyruvate kinase 120, 122, *123*

quin 2 25, 46, 124
quinones 17, 27, 172

rabbit 101, 168
redox cycling 2
  bipyridilium 17
  quinones 17
renal cells *202*
renal failure 165
renal injury
  oxygen-derived free radicals 166
renal ischaemia 165–87, *170*, *171*, *172*, 178, 181
  calcium 165
  cold 165, 166, 169, 170, 172
  iron 165
  lipid peroxidation 166
  oxygen free radicals 165
  warm 165, 166, 169, 170
renin release 178
reperfusion induced fibrillation
  protection by catalase 80
  protection by desferrioxamine 80
  protection by glutathione 80
  protection by mannitol 80
  protection by superoxide dismutase 80
reperfusion injury
  in kidney 166
reoxygenation
  cardiac tissue 140
respiratory burst 48
  cyanide 35
  NADPH oxidase 36
  phagocytosis 35
rose bengal 81, 82, *83*, *87*
  arrhythmias 81
  calcium current 85
Rubicon Hypothesis 189–212
ruthenium red
  mitochondrial calcium transport 174
ryanodine receptor 125, *126*, 128, 129, 131
  ryanodine gene 129

sarcolemma 125
  effect of cytosolic proteins 97

sarcolemma (*cont.*)
  molecular perturbation 100
  resting potential 97
sarcoplasmic reticulum 2, 125
  caffeine 81
  calcium release 82, 89, *90*
  damage 89
sarcotubular system
  lysosomes 159
  lysosomotropic agents 150
  swelling 150
selenium 119
  deficiency myopathy 140
skeletal muscle
  ultrastructural studies 149
sodium 100
sodium/calcium exchange *86*, 88, 89, *90*, 100, 102, *103*, 109
  free radicals 87
  in neutrophils 108
sodium influx 100, 102, *103*, 108
sodium pump
  current 85
  inhibition 87, *90*
  rose bengal *86*
spin labelling 124
staurosporine *47*, 48, 51
sulphydryl groups 101, 109
  electron acceptors 109
superoxide 11, 12, 17, 25, 60, 78, 108, 118, 172, 175, 181, 193, 196, *197*, 198, 199, 202
  ischaemia and reperfusion 78
  neutrophils 109
superoxide dismutase 2, 13, *19*, 78, 79, 80, 98, 101, 106, 166, 182
  copper/zinc-containing form 119
  manganese-containing form 119
superoxide radicals 166
surfactant secretion 161
synoviocytes *202*, 205

T lymphocyte *197*
T-tubules *157*
  endocytosis 159
12-*O*-tetradecanoyl-phorbol-13-acetate (TPA) 25, 26, 102, 106, 108
tetraethylammonium 84
thiobarbituric acid reactive substances (TBARS) 120, 121, *121*, *178*
  hydrogen-peroxide-induced 123
thiol homeostasis 27
thiol oxidation
  microfilament organisation 20
α-tocopherol *see* vitamin E
thromboxanes 175, *181*
tonicity 108
tributyltin 24
troponin C 124, 125, 193
type II cells 161
  alveolar epithelial cells 160
  endoplasmic reticulum 160
  inclusion bodies 161
  secretion of surfactant 150
tyrode solution *84*, *86*

ubiquitin-dependent proteases 23
uric acid 167

ventricular fibrillation 109
verapamil
  kidney 174
vitamin C 119
vitamin E 13, 14, 119, 120, *121*, 166, 176
  antioxidant 142
  lactate dehydrogenase release 143
  membrane permeability to Ca^{2+} 120
  supplementation 122, 142
vitamin E deficient 159
  exercise 142

xanthine dehydrogenase
  calcium-dependent proteolysis 167
  conversion to xanthine oxidase 167, 174
xanthine oxidase 11, 17, 78
  conversion from xanthine dehydrogenase 79, 167, 174

Z-lines 149
  blurred 149, *154*, *155*
zinc 24